OXFORD SCIENCE P

Mathematics Masterclasses

Stretching the imagination

Question and answer

Mathematics Masterclasses
Stretching the imagination

Edited by

MICHAEL SEWELL

Professor of Applied Mathematics, University of Reading

OXFORD   NEW YORK   TOKYO

OXFORD UNIVERSITY PRESS

1997

Oxford University Press, Great Clarendon Street, Oxford OX2 6DP

Oxford New York
Athens Auckland Bangkok Bogota Bombay Buenos Aires
Calcutta Cape Town Dar es Salaam Delhi Florence Hong Kong
Istanbul Karachi Kuala Lumpur Madras Madrid Melbourne
Mexico City Nairobi Paris Singapore Taipei Tokyo Toronto

and associated companies in
Berlin Ibadan

Oxford is a trade mark of Oxford University Press

Published in the United States
by Oxford University Press Inc., New York

A catalogue record for this book is available from the British Library

Library of Congress Cataloging in Publication Data
Mathematics Masterclasses—Stretching the imagination / edited by Michael Sewell.
1. Mathematics I. Sewell, M. J. (Michael J.), 1934–
QA3.M363 1997   510–dc20   96-27450
ISBN 0 19 851494 8 (Hbk)   ISBN 0 19 851493 X (Pbk)

Typeset by Technical Typesetting Ireland
Printed in Great Britain by Bookcraft (Bath) Ltd.
Midsomer Norton, Avon

Foreword

By GEORGE PORTER

Centre for Photomolecular Sciences, Imperial College

From 1975 to 1980 I was the first president of the National Association for Gifted Children, which sought to help bright young people who needed more support and encouragement than was available to those of average ability. The emphasis was heavily upon the arts, especially music, where it was recognised by everybody that education and learning could, and indeed should, start at a very early age. In mathematics and the sciences, education in the primary schools was limited, although many of us argued that children are born scientists.

In seeking to extend the work on behalf of the NAGC I noted that, whilst lectures in the sciences 'adapted to a juvenile Auditory', as the Royal Institution quaintly described them, had been popular and available to a fortunate few for over a century, mathematics was notably absent. This might have been due to the fact that the great scientist and teacher who founded these lectures, Michael Faraday, had almost no mathematics himself. Although, in his late teens, he was able to teach himself the experimental sciences, that was too late for him to begin mathematics.

Mathematics has much in common with music in that it can be practised and enjoyed by the very young. It is well known how Mozart and Menuhin were child prodigies whilst Maxwell and Newton were brilliant mathematicians at a similar age. Today one often hears of young people taking A-levels in mathematics in their early teens and graduating before they are twenty. But one usually finds that there is a tutor or parent devoted to the task of inspiring and teaching them. Those not blessed with such a mentor may be lost before their abilities have become apparent, even to themselves.

I chose mathematics, when initiating masterclasses at the Royal Institution, because of this analogy with music and the fact that talent in both areas is more easily recognised at an early age. The success of the first maths masterclasses at the R.I. gradually became noticed in other places and there are now over fifty series of these 'Royal Institution Mathematics Masterclasses' in Britain with over two thousand five hundred pupils each year.

Both those who are planning new masterclasses, and those who have benefited from them, can be helped by the experience of others and this book describes the contributions of the talented teachers who contributed to the very successful masterclasses given by the Berkshire group. It is edited by Professor Sewell of Reading University who was the organizer and the inspiration for this series and for this book. It will be welcomed by other masterclasses in the network and by all young mathematicians. It is a fascinating introduction to the joys of mathematics for all, including those not primarily mathematicians. It describes a kind of mathematics class that encourages us to enjoy, from an early age, the geometry and numeracy of the natural world.

Preface

The idea of a masterclass has become a familiar one, certainly in music. This is a book of masterclasses in mathematics. To read it requires no more prerequisites than could be possessed by a 13-year-old who has a curiosity about the subject. Mathematics is a language for solving problems.

In 1826 Michael Faraday gave some Christmas Lectures for young people about science, at The Royal Institution of Great Britain in Albemarle Street, London, where he had his laboratory. This began an annual series, which has been delivered by famous scientists every year since then, except for a short break during the Second World War. Faraday himself gave the lectures nineteen times up to 1860. The topic is usually selected from the experimental sciences, such as astronomy, biology, chemistry, medicine or physics. The lectures have been transcribed into popular book form some fifty times. Faraday's *The Chemical History of a Candle*, and Jean's *Through Space and Time* are famous examples. In 1978 mathematics was chosen as the subject for the first time, and Christopher Zeeman gave six one-hour lectures about '*The Nature of Mathematics and the Mathematics of Nature*', with demonstrations and illustrations.

The success of those lectures prompted George Porter, then Director of The Royal Institution, to promote the idea of mathematics masterclasses for young people. Zeeman gave the first such masterclasses in London in 1981, on spherical and perspective geometry. Enthusiasts in other venues then began to organize their own local groups, either in the bigger cities or on a county-wide basis, with the backing of The Royal Institution supplemented by local sponsorship. John Crank, as Chairman of the Mathematics Masterclasses Advisory Board at The Royal Institution until 1988, was a driving force in that initiation and development phase. There are now over 30 such groups, and each aims to provide one or more series of masterclasses annually for boys and girls who are aged 13 or not much more. This is the age at which, as some people believe, the way in which abstraction arising in mathematics can first begin to be appreciated.

Mathematics masterclasses for Berkshire began in 1991, and in each January and February since then we have provided two simultaneous series on six successive Saturday mornings. Each masterclass lasts $2\frac{1}{2}$ hours, divided into three sessions of exposition by the speaker, interspersed with two sessions of questions being worked by participants with help, if needed, from volunteer school teachers. This book is a selection from the 23 different masterclasses that we have arranged in six years, but transcribed into a form more appropriate to the written word.

A masterclass will succeed if it is both testing and enjoyable. Taken together these are, in my opinion, the crucial criteria, and pupils' and parents' written reactions afterwards strongly support that view. One parent said that his daughter 'returned happy and stimulated' and her participation 'may have far-reaching effects on her ambition and vision of the future'. The material here is not, however, confined to 13-year-olds. Older children can benefit, perhaps also with light guidance, and so can interested adults such as that father who said to me, 'I wish that maths had been like that when I was a lad'. The subject is also an international one, and our material is equally accessible to those abroad as to those in the United Kingdom.

Masterclass structure contains pauses for consolidation, and the following pause in this Preface reflects that feature.

Preface Worksheet Question: The Masterclass Organizer's Problem
The organizer has received a applications to attend a series of masterclasses to be held in a room where there are p places, and $2p \geq a > p$. There are an even number of masterclasses in the series. The organizer decides to solve the problem by inviting x applicants to come to the whole series, $\frac{1}{2}y$ applicants to come to half the series, and another $\frac{1}{2}y$ applicants to come to the other half of the series. Work out how each of x and y depend on a and p, and illustrate your method graphically. The solution is given at the end of this Preface.

The inequality $a > p$ summarizes the common experience that masterclasses are always oversubscribed. This gratifying feature expresses not only the natural ambition of parents and teachers for their charges, but also the intrinsic appeal of mathematical ideas to inquisitive young people. When the number of schools, s, does not exceed the number of places, the inequality $2p \geq a$ is a consequence of an expedient restriction that no school can nominate more than two applicants ($s \leq p$ and $a \leq 2s$ implies $a \leq 2s \leq 2p$). I have been very aware that, among the $60 +$ schools in our area which regularly send participants, a good number could find several pupils who are able or keen enough to benefit.

The presence of so few prerequisites might seem, at first, to be a serious restriction on the number of different topics which could successfully be treated in a sustained masterclass. For example, the idea of a limit, and therefore differentiation in particular, is not available and so cannot be used without specific preparation. Experience shows, however, that masterclasses have been constructed on very many topics, and new titles appear annually in a newsletter produced by The Royal Institution. The construction of a masterclass turns out to be an opportunity to develop an innovative approach to a piece of mathematics, either pure or applied. Such material frequently does not appear in current school syllabuses. The topic can be a classical one, for example in geometry; but it can also be one derived from relatively recent research work, and this book contains some of that. The speaker may be a school teacher, an industrialist or a member of a research establishment, or a university lecturer. The latter may know the boundary of knowledge in a research area and, because not all research requires difficult mathematics, he or she can quickly give some modern ideas a wider currency. This encourages new growth in the subject. There is a sense in which a masterclass can be 'elementary and advanced at the same time', as an American friend said to me after perusing a text. Masterclasses reveal a variety of styles of exposition from the different speakers, and I hope this book reflects that reality.

There is growing concern in England and Wales that school syllabus changes over the last two or three decades have had some very serious side effects on the mathematical preparedness of first-year undergraduates in mathematics, science and engineering. The London Mathematical Society, the Institute of Mathematics and its Applications, and the Royal Statistical Society instituted a joint working party to study this problem. In their 1995 report, *Tackling the Mathematics Problem*, they highlighted three particular deficiencies: (a) serious lack of manipulative ability; (b) lack of analytical confidence to address two-step and multi-step problems; (c) no understanding of the idea of proof or of the role of logical exposition in mathematics. This is a frightening indictment, bearing in mind how it might propagate to the next generation.

The speaker in a masterclass has a brief opportunity to draw the attention, at the very least, of a significant audience to the importance of those three points, and this volume may help in that way. In particular, the format of theorem and proof can be adopted in certain places, in both pure and applied mathematics. Some of the masterclasses in this book do that. In a recent masterclass I enquired how many participants could say what QED meant. No one could. The chance is also there to explain how real mathematics advances in an untidy and intuitive way, via hunches and guesses, and quite differently from the final logically expressed outcome, so that one should not be afraid to guess and test along the way. In 1992 a parent said 'the country needs to have its mathematical talent encouraged and developed in such a friendly and positive way'. Masterclass participants not infrequently say words to the effect that they have 'discovered a far more interesting side to maths than I have come across in school'. The publication of these and further masterclass texts may provide fresh resource material helpful to those who are in a position to promote mathematics at pivotal points in the education of young people, both in the United Kingdom and abroad. The chapters in this volume can be read in any order, and each is self-contained, not depending on any other.

Solution to the Masterclasses Organizer's Problem

The space restriction is

$$x + \tfrac{1}{2}y = p.$$

The pressure of applications is

$$x + y = a.$$

This is a pair of simultaneous linear equations, each of which has a straight line graph, as shown when $2p > a > p$. Their solution is where the straight

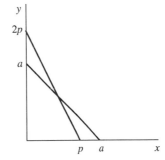

lines cross. By subtraction and back-substitution, the coordinates of that point are

$$x = 2p - a, \qquad y = 2(a - p).$$

Reading   Michael Sewell
November 1996

Acknowledgements

An especially rewarding feature of working with mathematics masterclasses is that everyone who one meets in that connection is an enthusiast. I have been indebted to many people. Having previously taught only undergraduates, I doubted at first whether I possessed the facility to give a masterclass. I learnt how to write one by studying Sir Christopher Zeeman's notes on *Geometry and Perspective* and on *Gyroscopes and Boomerangs*, which were produced as booklets by The Royal Institution. He chose the former topic when he gave the first masterclass in the opening Berkshire series in 1991. I also thank him for helpful practical advice on the organization of masterclasses.

As Local Organizer for Berkshire, I have received unstinting help from a small Organizing Committee: Jean Head, Inspector for Mathematics at Shire Hall; Geoffrey Smith of Downe House School (since 1992, and Michael Hosking of Leighton Park School before him); and David Harries (since 1992, John Wright before him). In particular, Jean and Geoff (or Michael) have collected nominations from the L.E.A. and the Independent schools respectively, and coordinated a rota of volunteer teachers. With each of our two classes of 36 pupils we have frequently had nine or more teachers on hand, giving their free time on Saturday mornings to help during the exercise and discussion sessions. That commitment speaks for itself.

The masterclasses have been held in the Department of Mathematics at the University of Reading. Three of our secretaries, Brigitte Calderon, Sue Davis and (until 1993) Joyce Bird have worked, not only steadfastly but enthusiastically, to make things actually happen when they should happen, and it would not have been possible to manage without them. Brigitte has also done many things which have been necessary to gather together the manuscript for this book. To them I am very grateful, and so, I believe, would be the participants if they could know what goes on behind the scenes.

The firm support of The Royal Institution has been with us throughout and, in particular, I have appreciated the encouragement expressed by Roger Bray, their Clothworkers' Fellow in Mathematics. Each local group is required to find, if possible, its own finance. I am pleased to acknowledge grants from Thames Valley Enterprise, which is our region's Training and Enterprise Council, for each of the four years 1992–5, and from the London Mathematical Society for 1996. They have made our initiative viable.

This volume contains only half of the masterclasses which have been given at Reading in six years. I regret that there is no space for more, but I hope those speakers whose texts are not represented on this occasion will be aware, nevertheless, of appreciation from the organizers and participants for their efforts. It is a pleasure to thank the contributors themselves for responding so willingly to my various requests.

The positive comments of parents each year, either in writing or at our closing sherry party, have encouraged us greatly. It is they who have been among the first to notice, for example, that the content of school syllabuses has been failing to provide appropriate stimulus. Every year they ask if there could be 'follow-up' in some form. There is clearly a substantial need for that, but we have been unable to provide it thus far on the voluntary basis in which masterclasses have worked up to now.

A masterclass audience might be the brightest group that a speaker ever encounters. Our speakers will wish me to thank the participants themselves, for reminding us each year of the liveliness and promise of each new generation. That is a feature which has been noticed by the local press, radio and television. In this connection I am pleased to acknowledge the assiduous work of Kay Dickinson, Information Officer at the University of Reading.

The percipient response of Elizabeth Johnston and her colleagues at Oxford University Press has been most welcome, and I have enjoyed working with them. I acknowledge permission from Thames Valley Newspapers to publish the photograph in the frontispiece, which first appeared in their *Evening Post* on 22 January 1993. It shows Patrick Evans of Cheam School, Newbury, asking a question of Roger Loveys of Highdown School, Reading.

M.J.S.

Contributors

Mr C. J. Bills, Department of Mathematics, Biddulph High School, Staffordshire.

Mrs E. J. Bills, School of Education, Crewe and Alsager Faculty, Manchester Metropolitan University.

Dr F. W. Dalton, Department of Mathematics, Harrow School.

Mr D. Harries, Department of Science and Technology Education, University of Reading.

Professor B. J. Hoskins, F.R.S., Department of Meteorology, University of Reading.

Dr F. C. Kirwan, Balliol College, University of Oxford.

Dr P. K. Maini, Centre for Mathematical Biology, University of Oxford.

Professor D. J. Needham, Department of Mathematics, University of Reading.

Professor Lord Porter, O.M., F.R.S., Centre for Photomolecular Sciences, Imperial College.

Professor M. J. Sewell, Department of Mathematics, University of Reading.

Dr D. S. G. Stirling, Department of Mathematics, University of Reading.

Dr A. A. White, The Meteorological Office College, Reading.

Dr W. L. Wood, Department of Mathematics, University of Reading.

Contents

1 A Little Bit of Chaos

by CHRIS BILLS
Biddulph High School, Staffordshire

and LIZ BILLS
Crewe and Alsager Faculty, Manchester Metropolitan University

1. Introduction

You may have seen pictures something like this before. They are often called 'fractals' or 'chaos theory' pictures. Figure 1 shows two *Julia sets* and Fig. 2 is the *Mandelbrot set*. Both are named after mathematicians who were among the first to work on them. *Gaston Julia* was born in 1893 and died in 1978. He did his first work on the sets now known as Julia sets in 1916 when he was recovering in hospital. He had been injured as a result of the First World War. The Mandelbrot set is named after *Benoit Mandelbrot* who is still working as a mathematician today. It was first produced in 1979, so compared with most of the mathematics you learn about in school it is a very recent discovery.

Fig. 1.　Julia sets

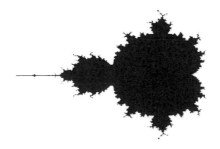

Fig. 2.　Mandelbrot set

These sets have been studied by mathematicians interested in *Chaos* because they are very complex images resulting from very simple calculations. Another chapter in this book, 'Discrete Mathematics and its Application to Ecology', introduces you to another aspect of Chaos. In this chapter we intend to show you the basic ideas behind Julia sets and the Mandelbrot set. To do that we will have to start with some things which seem to have nothing to do with these pictures.

2. Complex numbers

Can you solve the equation

$$x^2 - 9 = 0?$$

There are two solutions 3 and -3. What about

$$x^2 - 1 = 0?$$

Two solutions again. If you have learnt about quadratic equations (equations with a square as the highest power) then you will know that they often have two solutions. What about this one?

$$x^2 + 1 = 0$$

Amongst the *real* numbers there are no solutions to this equation. There is no real number whose square is minus one. In fact for *any* negative number there is no real number which is its square root. Around the beginning of the nineteenth century mathematicians began using the symbol i to stand for the square root of minus one. They called the square roots of negative numbers *imaginary* numbers.

So using this notation

$$i^2 = -1,$$
$$(3i)^2 = -9,$$
$$(-2i)^2 = -4 \quad \text{etc.}$$

Also

$$\sqrt{(-4)} = \pm 2i,$$
$$\sqrt{(-121)} = \pm 11i \quad \text{etc.}$$

Using these imaginary numbers mathematicians were able to solve a lot of equations which they could not solve with real numbers. But there were a lot more which needed something more. For example to solve the equation

$$x^2 - 2x + 2 = 0$$

we need a real number and an imaginary number added together. The two solutions to this equation are $1 + i$ and $1 - i$. (You are asked to check this in Worksheet 1.) Numbers like this are called *complex numbers*.

Complex numbers can be added, subtracted, multiplied and divided. For our purposes in this chapter we need to know how to add and multiply. Here are two examples of adding:

$$2 + 3i + 1 - 4i = 2 + 1 + 3i - 4i$$
$$= 3 - i$$
$$-1 + 4i - (-2 + 3i) = -1 + 2 + 4i - 3i$$
$$= 1 + i$$

and two of multiplying

$$(2 + i)(3 + 4i) = 2 \times 3 + 2 \times 4i + i \times 3 + i \times 4i$$
$$= 6 + 8i + 3i + 4i^2$$
$$= 6 + 11i + 4 \times (-1)$$
$$= 6 + 11i - 4$$
$$= 2 + 11i$$
$$(-1 + 2i)(0.5 - i) = -1 \times 0.5 + (-1) \times (-i) + 2i \times 0.5 + 2i \times (-i)$$
$$= -0.5 + i + i + (-2i^2)$$
$$= -0.5 + 2i + (-2) \times (-1)$$
$$= -0.5 + 2i + 2$$
$$= 1.5 + 2i$$

The process we will need most is squaring a complex number, that is multiplying it by itself. Here is an example:

$$(3 - i)^2 = (3 - i)(3 - i)$$
$$= 3 \times 3 + 3 \times (-i) + 3 \times (-i) + (-i) \times (-i)$$
$$= 9 + (-3i) + (-3i) + i^2$$
$$= 9 - 6i - 1$$
$$= 8 - 6i$$

Worksheet 1

1. Find the solutions of these equations.
   (a) $x^2 - 4 = 0$
   (b) $x^2 + 4 = 0$
   (c) $3x^2 = 48$
   (d) $-50 = 2x^2$
2. Work out
   (a) $(2i)^2$
   (b) $\sqrt{(-36)}$
   (c) $\sqrt{(-0.04)}$
   (d) $(0.3i)^2$
3. Work out
   (a) $(1 + i)(1 + 2i)$
   (b) $(0.2 - 1.5i)(-0.4 + 1.2i)$
   (c) $(3 + 2i)^2$
   (d) $(0.2 + 0.3i)^2$

4. Check that $1 + i$ and $1 - i$ are solutions of $x^2 - 2x + 2 = 0$.

5. Work out $(x + yi)^2$.

6. Work out $(1 + i)(1 - i)$, $(2 + 3i)(2 - 3i)$, $(0.2 + 0.3i)(0.2 - 0.3i)$. What do you notice? Can you give a general rule? Why does it happen?

Solutions to the worksheet are given at the end of the chapter.

3. The Argand diagram

Complex numbers are often represented on a grid called an Argand diagram. (It is named after the first mathematician to use it.) The real part of the complex number tells you how far to go right or left and the imaginary part tells you how far to go up or down, rather like plotting coordinate pairs. In Fig. 3 some complex numbers are plotted on an Argand diagram to show you how it works.

We can measure the *size* of a complex number by calculating its distance from $(0, 0)$ on the Argand diagram. To do this you need to use Pythagoras' rule about right-angled triangles. Figure 4 shows the following example.

The distance of $3 + 2i$ from $(0, 0)$ is the same as the length of the line AB. But from the triangle ABC, using Pythagoras' rule

$$AB^2 = AC^2 + BC^2 = 3^2 + 2^2 = 9 + 4 = 13$$
$$AB = \sqrt{13}$$

4. Iterations

Here is a rule, called a *mapping*, which we could apply to a complex number:
$$z \to 2z + 3.$$

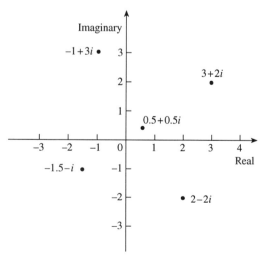

Fig. 3.   Complex numbers on an Argand diagram

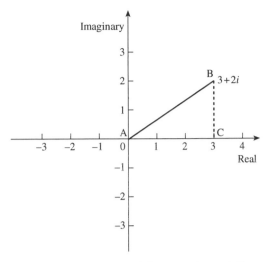

Fig. 4.   Distances from the origin on an Argand diagram

(Complex numbers are often called z, just as real numbers are often called x). This rule tells me to double z and add 3. The arrow is read as *maps onto*. For example if we apply the mapping to $1 + i$ we get

$$2(1 + i) + 3 = 5 + 2i.$$

Then if we apply it to $5 + 2i$, we get

$$2(5 + 2i) + 3 = 13 + 4i.$$

If we continue by applying the mapping to $13 + 4i$ we are performing an iteration. That just means we are using the same mapping over and over, each time on the result of the last application.

The mapping we are going to look more closely at is

$$z \rightarrow z^2$$

Let's apply this several times, starting with $0.5 + 0.4i$

$$(0.5 + 0.4i)^2 = 0.5 \times 0.5 + 0.5 \times 0.4i + 0.5 \times 0.4i + 0.4i \times 0.4i$$
$$= 0.25 + 0.2i + 0.2i + 0.16i^2$$
$$= 0.25 + 0.4i + 0.16 \times (-1)$$
$$= 0.09 + 0.4i$$
$$(0.09 + 0.4i)^2 = \cdots = -0.1519 + 0.072i$$
$$(-0.1519 + 0.072i)^2 = 0.017\,889\,61 - 0.021\,873\,6i$$
$$(0.017\,889\,61 - 0.021\,873\,6i)^2 \approx -0.000\,158\,4 - 0.000\,782\,6i$$
$$(-0.000\,158\,4 - 0.000\,782\,6i)^2 \approx -0.000\,000\,5 + 0.000\,000\,2i$$

(In the last two equations the $=$ sign is replaced by $\approx$ because the answers given are only approximate.)

If we continue with this iteration the real and imaginary parts of the complex number continue getting closer and closer to zero. On a calculator the very small real and imaginary parts would be written in *standard form*. For instance the last imaginary part we have calculated, 0.0000002, would appear on the calculator as

$$2.0\text{E} - 07.$$

Large numbers may also appear on the calculator in standard form. For instance $23\,600\,000\,000$ would appear as $2.36\text{E}10$.

If we plot this sequence of complex numbers on an Argand diagram the points will get closer and closer to $(0,0)$.

So starting from the complex number $0.5 + 0.4i$ and applying the iteration $z \to z^2$ we get a sequence of complex numbers which get closer and closer to zero.

Now let's try $1 + i$ as our starting number.

$$(1+i)^2 = 1 \times 1 + 1 \times i + 1 \times i + i \times i$$
$$= 1 + i + i + (-1)$$
$$= 2i$$
$$(2i)^2 = 4i^2 = -4$$
$$(-4)^2 = 16$$
$$16^2 = 256$$

And so on.

You can see that the numbers in this sequence are going to keep getting bigger as we continue to iterate. By this we mean that they are going to continue to get further away from $(0,0)$.

On a piece of graph paper plot an Argand diagram like the one in Fig. 3 but make the real and imaginary axes each go from minus ten to ten. Plot points to represent the first few numbers which we got by applying the iteration $z \to z^2$ to $1 + i$. Plot as many as you can of the points we got when we applied the iteration to $0.5 + 0.4i$ as well.

Choose a third complex number for yourself and apply the iteration to it. Plot a point on your Argand diagram each time you get a new answer. Do your points keep getting closer to $(0,0)$ or further away?

5. Prisoners and escapers

By now we will have seen two classes of complex numbers, those whose iterated sequences contain numbers which *tend to zero* and those whose iterated sequences contain numbers which continue getting further from zero. The first set are called *prisoners* and the second set are called *escapers*. In fact for this iteration, almost all starting points give sequences in which the numbers either tend to zero or get further and further away from zero. In other words almost all points are either prisoners or escapers. There are a few points which are exceptions to this. We will talk about them again at the end of Worksheet 2.

The calculations to find out whether a point is a prisoner or an escaper are quite time-consuming, so we have written a graphic calculator program which will do the calculations for you. The programs are written for Casio and Texas Instruments calculators but it is possible to do something very similar on any other model. This program decides whether a point which you choose is a prisoner or escaper for the iteration $z \to z^2$. Here are listings of the programs.

CASIO	TEXAS INSTRUMENTS
"FIRST X"? $\to$ X	Input "FIRST X ?", X
"Y"? $\to$ Y	Input "Y?", Y
1 $\to$ N	1 $\to$ N
Lbl 1	Lbl 1
$X^2 - Y^2 \to D$	$X^2 - Y^2 \to D$
2XY $\to$ Y	2XY $\to$ Y
D ◢	Disp D
Y ◢	Disp Y
" "	Pause
D $\to$ X	D $\to$ X
N+1 $\to$ N	N+1 $\to$ N
N > 20 $\Rightarrow$ Goto 2	If N > 20
$X^2 + Y^2 < 100 \Rightarrow$ Goto 1	Goto 2
"ESCAPER"	If $X^2 + Y^2 < 100$
Goto 3	Goto 1
Lbl 2	Disp "ESCAPER"
"PRISONER"	Goto 3
Lbl 3	Lbl 2
	Disp "PRISONER"
	Lbl 3

If you did question 5 on Worksheet 1 you probably recognize that the important lines in this program are

$$X^2 - Y^2 \to D \quad \text{and}$$
$$2XY \to Y$$

because these are the lines where the real and imaginary parts of new complex number are calculated. Each time the new complex number is calculated the program checks that it is not yet too big using the lines

$$\text{If } X^2 + Y^2 < 100$$
$$\text{Goto 1}$$

or

$$X^2 + Y^2 < 100 \Rightarrow \text{Goto 1}.$$

(Here we have said that a complex number is too big if it is more than ten units away from zero. In doing this we are saying that once it has got as big as this we are confident that it will carry on getting larger. The choice of ten is not quite arbitrary and we will say a bit more about this later.)

If it is very large the program returns the message 'ESCAPER'. If it is not yet too big the calculator returns to Lbl 1 and performs the next iteration. Each time this happens N is increased by 1 so that the calculator counts how many times the iteration has been performed. If the number is still not too big after twenty iterations the program returns the message 'PRISONER'. We could be more convinced that we had found all the prisoners if we increased the number of iterations from twenty. On the other hand, the program would then take longer to run. So twenty is a compromise.

To use the program enter the real and imaginary parts of your starting number at the prompts 'FIRST X?' and 'Y?'. The calculator displays the real and imaginary parts of each new complex number as the iteration goes on. Keep pressing EXE or ENTER until you see either 'PRISONER' or 'ESCAPER' on the screen.

Worksheet 2

In this worksheet we are going to suggest how you might use the calculator program to investigate which points are prisoners and which are escapers for the iteration $z \to z^2$.

1. Draw out another Argand diagram on graph paper. This time make each axis go from minus two to two. Use as large a scale as possible.
2. Choose a complex number to start the iteration and enter the real and imaginary parts into the calculator. Run the program.
3. When you have found whether the number you chose is a prisoner or an escaper mark it on to your diagram with a colour coding – for example, black for a prisoner and red for an escaper.
4. Try several different complex numbers from different areas of the Argand diagram as your starting points. Mark each in red or black as you find whether it is a prisoner or an escaper. You should begin to see a pattern emerging as you place more and more points on your diagram.

Comments on Worksheet 2

You will find that any points whose distance from zero is less than one are prisoners. The reason for this is that when a complex number is squared, its distance from zero is also squared. So if its distance from zero is less than one, when it is squared this distance gets smaller, that is it gets closer to zero. On the other hand, if its distance from zero is greater than one, when it is squared the distance gets larger. So, on the Argand diagram, any points which lie within a circle of radius one with its centre at the origin are prisoners. These points form the *prisoner set*. If we draw a diagram to show the prisoner set it will show a circle.

Points inside the circle are prisoners, and tend to zero. Points outside the circle are escapers and become very large as they are iterated. What happens to points which lie *on* the circle? The calculator program is not very helpful here because it is not designed to deal with points on the circle. In addition the calculations are not entirely accurate and the inaccuracies build up after a few iterations. However, we can decide what happens to these points by thinking about their distance from zero. A point which lies on the circle is a distance of exactly one from zero. So when it is squared, its distance from zero remains one. In other words it stays on the circle. It may move to another point on the circle but it cannot move off it, either towards zero or away from zero. Try looking at the behaviour of these points, all of which are on the circle, as you perform the iteration on them

$$1, -1, i, -i, 0.8 + 0.6i$$

6. Julia sets

The name *Julia sets* is applied to the boundaries of the prisoner sets for certain iterations. The iterations are all of the form

$$z \to z^2 + c$$

where c can be any complex number. The two which you saw in Fig. 1 were for $c = -0.5 + 0.5i$ and $c = -1$. The Julia set for $c = 0$ is a circle, as we have just seen.

It is possible to adapt the graphic calculator programs so that they test whether a point is a prisoner or escaper for any iteration you choose of the form $z \to z^2 + c$. The adapted programs are listed in Appendix A. The new programs still contain the lines

$$\text{If } X^2 + Y^2 < 100$$
$$\text{Goto } 1$$

or

$$X^2 + Y^2 < 100 \Rightarrow \text{Goto } 1.$$

So we are still testing whether the number is further than ten away from zero. In fact it is possible to show that for any particular value of c, we only need to test whether it has got further than $\frac{1}{2} + \sqrt{(\frac{1}{4} + |c|)}$ from zero.

However we decided not to overcomplicate the program by setting a threshold which depends on the value of c. Ten is big enough for most interesting values of c, and not so big that it slows the program down too much.

A computer program can very quickly check if any point on the screen (representing the Argand diagram) is a prisoner or not and plot them in a different colour to escapers. A program written in BBC basic is given in Appendix B. The important lines in this program are

```
160   X(K+1)=X(K)^2-Y(K)^2+R
170   Y(K+1)=2*X(K)*Y(K)+I
```

X(K+1) and Y(K+1) are the real and imaginary parts of the new complex number. R and I are the real and imaginary parts of c.

The computer checks the size of the complex number using the lines

$$140 \quad D=SQR(X(K)^2+Y(K)^2)$$
$$150 \quad IF \ D>10 \ GOTO \ 200$$

Escapers are plotted in white and prisoners are left black. The machine then starts with a new starting point.

The program we have suggested uses the simplest possible approach. There are other methods of producing these drawings more efficiently. You will find details of some of these in the books listed under 'Further reading'.

7. The Mandelbrot set

We now turn our attention to the Mandelbrot set. This is not formed in the same way as the Julia sets, but is one stage more complicated. To form each Julia set we have to choose a value of c to put in the iteration $z \rightarrow z^2 + c$. Then the Julia set is the boundary of the set of points which are prisoners for this iteration. Some of the prisoner sets formed in this way contain zero and some of them do not. The values of c which give prisoner sets which contain zero are members of the Mandelbrot set. In other words the Mandelbrot set contains all those values of c for which the prisoner set contains zero. It can also be shown that any Julia set for which the value of c is an element of the Mandelbrot set is *connected*. A set is connected if I could travel from any point in the set to any other without going outside the set. For example, in Fig. 5, sets (i), (iii), (iv) and (v) are connected whereas (ii) and (vi) are not. That means that the complex numbers $-0.71 + 0.36i$, -0.5, $-0.05 + 0.745i$ and $0.27 + 0.49i$ lie inside the Mandelbrot set whilst 0.3 and $-0.73 + 0.38i$ lie outside.

The program we have used to generate the Mandelbrot set is listed in Appendix C. This program chooses its own values for c and checks if zero is a prisoner. The lines

$$90 \quad X(0)=0$$
$$100 \quad Y(0)=0$$

start each iteration from $(0,0)$. If zero is an escaper the program plots c in white. If not it leaves it black.

The two programs listed in Appendices B and C each have to overcome the problem of the screen origin being in the bottom left-hand corner rather than the centre. Each uses a different approach to this problem.

Figure 5 shows some pictures which were produced using these programs. If you have access to a computer which runs Basic you can experiment with producing some of your own. If you have a little knowledge of programming you could improve on our programs by increasing the efficiency or adding colour. You might also try writing a routine to zoom in on selected parts of the screen.

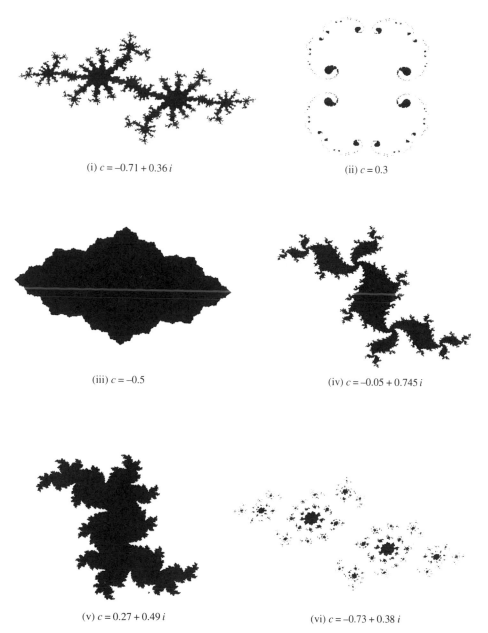

(i) $c = -0.71 + 0.36\,i$

(ii) $c = 0.3$

(iii) $c = -0.5$

(iv) $c = -0.05 + 0.745\,i$

(v) $c = 0.27 + 0.49\,i$

(vi) $c = -0.73 + 0.38\,i$

Fig. 5. More Julia sets

8. Conclusion

Julia sets and the Mandelbrot set are examples of *fractals*. The study of sets like these is called *fractal geometry*. Fractal geometry is an extension of classical geometry and has been used to construct models of physical phenomena such as mountains, coral reefs, coastlines and clouds. It has helped scientists to understand more about a very large number of natural structures. More recently it has found an application in image compression (recording a lot of detailed information about a screen image in a very small space), which turns out to be very important when you want to transmit an image from one computer to another, either on-line or by CD-ROM or floppy disk.

The key point in both of these areas of application is that a very complicated image results from the repeated application of a very simple rule. We hope that this chapter has helped you begin to understand how this might occur.

Solutions to Worksheet 1

1. (a) $x = 2$ or $x = -2$
   (b) $x = 2i$ or $x = -2i$
   (c) $x = 4$ or $x = -4$
   (d) $x = 5i$ or $x = -5i$
2. (a) -4
   (b) $6i$ or $-6i$
   (c) $0.2i$ or $-0.2i$
   (d) -0.09
3. (a) $(1 + i)(1 + 2i) = 1 \times 1 + 1 \times 2i + i \times 1 + i \times 2i$
   $$= 1 + 2i + i + (-2)$$
   $$= -1 + 3i$$
   (b) $(0.2 - 1.5i)(-0.4 + 1.2i) = 0.2 \times (-0.4) + 0.2 \times 1.2i + (-1.5i)$
   $$\times(-0.4) + (-1.5i) \times 1.2i$$
   $$= -0.08 + 0.24i + 0.6i + 1.8$$
   $$= 1.72 + 0.84i$$
   (c) $(3 + 2i)^2 = (3 + 2i)(3 + 2i)$
   $$= 3 \times 3 + 3 \times 2i + 2i \times 3 + 2i \times 2i$$
   $$= 9 + 6i + 6i - 4$$
   $$= 5 + 12i$$
   (d) $(0.2 + 0.3i)^2 = (0.2 + 0.3i)(0.2 + 0.3i)$
   $$= 0.2 \times 0.2 + 0.2 \times 0.3i + 0.3i \times 0.2 + 0.3i \times 0.3i$$
   $$= 0.04 + 0.06i + 0.06i - 0.09$$
   $$= -0.05 + 0.12i$$
4. $(1 + i)^2 = 2i$,
   so if $x = 1 + i$,

   $$x^2 - 2x + 2 = 2i - 2(1 + i) + 2$$
   $$= 2i - 2 - 2i + 2$$
   $$= 0$$

   $(1 - i)^2 = -2i$,

so if $x = 1 - i$,

$$x^2 - 2x + 2 = -2i - 2(1 - i) + 2$$
$$= -2i - 2 + 2i + 2$$
$$= 0$$

5. $(x + yi)^2 = x \times x + x \times yi + yi \times x + yi \times yi$
$$= x^2 - y^2 + 2xyi$$

6. $(1 + i)(1 - i) = 2$
$(2 + 3i)(2 - 3i) = 13$
$(0.2 + 0.3i)(0.2 - 0.3i) = 0.13$

The answer in each case is real, in other words the imaginary part of the answer is zero. In general a product of the form $(a + bi)(a - bi)$ is equal to $a^2 + b^2$. This happens because the two imaginary terms in the product are equal but opposite in sign.

Appendix A

Graphic calculator programs to test whether a point is an escaper or a prisoner for the iteration $z \to z^2 + c$.
Here are the listings of the programs:

```
CASINO                          TEXAS INSTRUMENTS

"A IN C=A+BI" ? → A              Input "A IN C=A+BI?", A
"B" ? → B                        Input "B?", B
"FIRST X" ? → X                  Input "FIRST X?", X
"Y" ? → Y                        Input "Y?", Y
1 → N                            1 → N
Lbl 1                            Lbl 1
X² − Y²+A → D                    X² − Y²+A → D
2XY+B → Y                        2XY+B → Y
D ◢                             Disp D
Y ◢                             Disp Y
" "                             Pause
D → X                            D → X
N+1 → N                          N+1 → N
N > 20 ⇒ Goto 2                  If N > 20
X²+Y² < 100 ⇒ Goto 1             Goto 2
"ESCAPER"                        If X²+Y² < 100
Goto 3                           Goto 1
Lbl 2                            Disp "ESCAPER"
"PRISONER"                       Goto 3
Lbl 3                            Lbl 2
                                Disp "PRISONER"
                                Lbl 3
```

Appendix B Julia set program

```
10      REM HIGH RESOLUTION JULIA SETS
20      MODE 0
30      CLS
40      INPUT "REAL PART", R
50      INPUT "IMAGINARY PART", I
60      CLS
70      DIM X(20)
80      DIM Y(20)
90      FOR X=0 TO 638 STEP 2
100     FOR Y=0 TO 1020 STEP 4
110     X(0)=X*1.5/640-1.5
120     Y(0)=Y*1.5/512-1.5
130     FOR K=0 TO 19
140     D=SQR(X(K)^2+Y(K)^2)
150     IF D>10 GOTO 200
160     X(K+1)=X(K)^2-Y(K)^2+R
170     Y(K+1)=2*X(K)*Y(K)+I
180     NEXT K
190     GOTO 220
200     PLOT 69, X, Y
210     PLOT 69, 1278-X, 1020-Y
220     NEXT Y
230     NEXT X
240     PRINT "Re="; R; " Im="; I
250     *SCREENSAVE FRACT1
```

Appendix C Mandelbrot set program

```
10      REM HIGH RESOLUTION MANDELBROT SET
20      MODE 0
30      VDU 29, 960; 512;
40      CLS
50      DIM X(20)
60      DIM Y(20)
70      FOR R=-960 TO 320 STEP 2
80      FOR I=-512 TO 512 STEP 4
90      X(0)=0
100     Y(0)=0
110     FOR K=0 TO 19
120     D=SQR(X(K)^2+Y(K)^2)
130     IF D>10 GOTO 180
140     X(K+1)=X(K)^2-Y(K)^2+R*1.5/640
150     Y(K+1)=2*X(K)*Y(K)+I*1.5/512
160     NEXT K
170     GOTO 190
180     PLOT 69, R, I
190     NEXT I
200     NEXT R
210     *SCREENSAVE MAND
```

Further reading

H.-O. Peitgen and P. H. Richter, *The Beauty of Fractals: Images of Complex Dynamical Systems* (Springer-Verlag, New York, 1986)

A lavishly illustrated book with brief explanations of the mathematics which lies behind the computer graphics displayed. Some Basic programs are listed for you to try.

H.-O. Peitgen, H. Jürgens, and D. Saupe, *Fractals for the Classroom. Part 1. Introduction to Fractals and Chaos* (Springer-Verlag, New York, 1992)

A very detailed introductory account. A school student is more likely to understand some of the detail of this book and its partner volume than most others on the subject.

H.-O. Peitgen, H. Jürgens, and D. Saupe, *Fractals for the Classroom. Part 2. Complex Systems and the Mandelbrot Set* (Springer-Verlag, New York, 1992)

Partner volume to the above.

M. Barnsley, *Fractals Everywhere* (Academic Press, Orlando, Florida, 1988)

A very well illustrated book which contains some mathematical explanations which will be accessible to school students but quickly becomes quite advanced. Worth a look for the pictures and pictorial explanations though!

H.-O. Peitgen and D. Saupe (eds.), *The Science of Fractal Images* (Springer-Verlag, New York, 1988)

This book has a very good first chapter which details some of the history of the science of fractal images in a fairly non-technical way. Other chapters are at a similar level to those in *Fractals Everywhere*. It also has some excellent pictures and some computer program listings.

2   Nightmare on Large Number Street

by BILL DALTON
Department of Mathematics, Harrow School

1. Introduction

Computers play a vitally important role in the lives of us all. When we are taking off or landing in an aeroplane, our safety is dependent on the efficient operation of several computer systems. In most areas of finance – taxation, banking, personal credit – information is stored in, and provided by, computers. In hospitals and in doctors' surgeries, computers are now commonplace. So it is reasonable to ask what are computers and just how reliable are they?

What are computers?

Although it is difficult to be certain when any story begins, one possible starting point to our story is a meeting that was held in Paris in August 1900. Every so often, many of the world's best mathematicians meet and talk about mathematics. There are lectures and discussion groups, but at each meeting there are a few lectures that are more important than all the others. These are given by the greatest mathematicians of the day and all present look forward to hearing what these distinguished mathematicians have to say. In 1900, the man who many thought was probably the world's greatest mathematician was a thirty-eight-year-old German mathematician called David Hilbert.

Hilbert's address to the 1900 meeting has become very famous. Many people expected him to say where be thought mathematics would be heading in the new century. Others thought he would describe some of the important new results he had proved since the last meeting. But he did neither of these things and what he did was really quite remarkable. He listed twenty-three problems that he could not solve. These have now become known as *Hilbert's Problems*. Some of these problems, he said, would be solved within a few years, but others would remain unsolved for the whole of the twentieth century.

Was Hilbert right? Up until 23 June 1993 only two remained to be solved. Then, on that day, at the Newton Institute in the University of Cambridge, Andrew Wiles produced a solution for the problem known as Fermat's Last Theorem. As I write, only one of Hilbert's twenty-three problems has still not been solved. This is known as the *Riemann Hypothesis*. It will be interesting to see if a solution for this problem can be found before the year 2000. We, however, shall be interested in the tenth of Hilbert's problems.

2. Hilbert's 10th problem

Look at the equations written below.

$$5x - 2y = 4$$

$$4x + 8y = 7$$

$$111x^3y + 16xy^2 - xy^4z^5 = 1$$

Each of these equations involves two or more unknowns. The first two equations have two unknowns x and y. The third has three: x, y, and z. The coefficients appearing in all three equations are integers (whole numbers).

A *solution* for an equation is a numerical value given to each of the unknowns in the equation so that when the numerical values are substituted for the unknowns, the left-hand side of the equation equals the right-hand side.

An equation is called *Diophantine* if we are looking for a solution consisting entirely of integers (whole numbers). Such equations are named after Diophantus, who lived in Alexandria in approximately AD 250. He was, as far as we know, the first person to consider problems involving such equations.

In his 10th problem, Hilbert asked: is there a *computational procedure* that will test an equation (in any number of unknowns) to determine whether or not it has a solution consisting entirely of integers? In fact, Hilbert's exact words were:

'Given a Diophantine equation with any number of unknown quantities and with integer coefficients, devise a process according to which it can be determined by a finite number of operations whether the equation is solvable in integers.'

The first equation has the solution $x = 2$, $y = 3$; $x = 4$, $y = 8$; $x = 6$, $y = 13$; and so on. So this equation has several pairs of solutions which are whole numbers.

The second equation cannot have a solution of whole numbers. If there was such a solution, substitution for x and y would make the left-hand side an even number while the right-hand side would be an odd number.

For the third equation, the problem is a bit more difficult.

Many mathematicians tried to solve Hilbert's 10th problem, but initially with little success. Then, it became clear to some that the real difficulty lay in the fact that nobody really knew what a *computational procedure* was. So mathematicians began to think about computational procedures.

What, exactly, is a computational procedure? It has to be a finite sequence of steps where each step is a precise instruction and where each step (after the first) follows from the one before without the need for human (or outside) intervention. It has to be completely mechanical. So there must be a *set of rules* which would tell us, at each step, precisely what must be done next. What could these rules be? Would it be possible to write them down? Could they be used to solve problems?

In 1936, an English mathematician, called Alan Turing, showed that the answers to the last two questions were 'Yes' and 'Yes'. In that year, he described what came to be known as a *Turing Machine*.

At this point, our interest in Hilbert's tenth problem begins to fade and we shall not pursue this problem any further. The problem served to introduce the idea of a computational procedure. Alan Turing showed that computational procedures and Turing Machines were one and the same. So to study computational procedures further, we shall look at Turing Machines. But before we leave it altogether, we describe a little more of the history of Hilbert's 10th problem. The first real progress towards a solution was made in 1950 by Martin Davis, an American mathematician (see Davis (1978)). Other mathematicians made contributions and suggestions but it was not until 1970 that the final part of the proof fell into place. It fell to Yuri Matyasevich, a 22-year-old Russian mathematician, to show that no computable procedure exists for deciding whether or not a given Diophantine equation has an integer solution. The story of the search for a proof is fascinating and details may be found in Devlin (1990).

3. A Turing Machine

Before describing Turing Machines in detail, we make two remarks. First, we shall assume that any calculation may be written out in a single straight line. For example, if we wanted to calculate 25×38, we would probably write

$$
\begin{array}{rl}
25 & \\
38 & \times \\
\hline
200 & \\
750 & + \\
\hline
950 &
\end{array}
$$

But we could equally well have written

$$25 \times 38 = 200 + 750 = 950$$

So instead of working 'downwards' as we tend to do when working in mathematics, we shall work 'horizontally'. In this way, the problem, the intermediate steps and the solution may all be written on a single line on a 'very long' piece of paper. This piece of paper, which we shall call a tape, is divided up into squares. So this problem could be written

| | | 2 | 5 | × | 3 | 8 | = | 2 | 0 | 0 | + | 7 | 5 | 0 | = | 9 | 5 | 0 | | |

But secondly, the only symbols that are allowed to appear on the tape are 0 and 1. This simplifies the language of a Turing Machine considerably and it is not as restrictive as might at first be thought. We shall write 3 as

·	·	·	0	0	0	1	1	1	0	0	0	0	0	·	·	·

while 5 becomes

0	0	1	1	1	1	1	0	0	0	0	0		

We will show later than *any* number may be written using only 0 and 1. But beware; the system used in this section for writing numbers is not the binary system described and used in Section 5.

Now we think about Turing Machines.

Remember: a computable procedure is a sequence of steps: the steps are executed strictly in order; each step must be a precise instruction and it is always clear what the machine (or the operator) must do next.

A Turing Machine may be thought of as a box through which a tape is fed, as shown in Fig. 1.

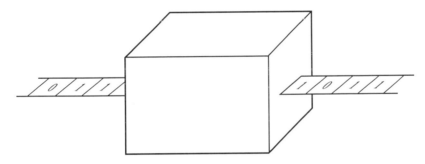

Fig. 1. Turing machine

At any moment, the machine is scanning a square (the square under consideration): the machine reads what is written in that square and responds accordingly. The rules that determine what the machine can do are the *set of rules* mentioned earlier. There are seven rules governing the behaviour of a Turing Machine and they are unbelievably simple. The seven rules are:

write 1
write 0
go right   (move one square to the right)
go left   (move one square to the left)
go to step *i* if 1 is scanned
go to step *j* if 0 is scanned
stop

Rules 5 and 6 might need a little explanation. In a Turing Machine, each step is one of these seven rules. So we might have:

1	write 0
2	go right
3	go to step 1 if 1 is scanned
4	stop

The order in which the steps are performed may not be altered. So we may number the steps (as we have done here). Now, step 3 says 'go to step 1 if 1 is scanned'. If there is a 1 in the square under consideration, then we go to step 1. Step 1 tells us to write 0 in that square. Then, we move to step 2. If, however, there is a 0 in the square under consideration then the instruction, 'go to step 1' is ignored and we proceed to step 4, and stop.

Alan Turing showed that, astonishingly, this set of rules is sufficient to calculate anything that ought to be *calculable*. Any computation whatsoever may be carried out using just these rules. They are what is meant by a *computable procedure* and the world's most powerful computers can do no more than is allowed by these seven simple rules.

In the following exercises, we look at how these rules may be used to carry out some simple computations.

Worksheet 1

In all these exercises, the machine is considering the square with the arrow directly above.

1. This tape contains a single 1.

↓							
0	0	1	0	0	0	0	0

What happens if the following sequence of instructions (*program*) is run?
go right
write 1
stop

2. The tape contains four ones.

	↓							
0	0	1	1	1	1	0	0	0	0

What happens if the instructions are:
1   go right
    go to step 1 if 1 is scanned
    write 1
    go right
    write 1
    stop

3. Start with a tape containing five ones.

0	0	1	1	1	1	1	0	0	0	0	0

Write a set of instructions that will calculate $5 - 2$.

How would your instructions change if you were calculating $n - 2$ where n was any positive integer?

4. A tape contains three ones.

0	0	1	1	1	0	0	0	0

Perform the following instructions. What has happened?

1   go right
     go to step 1 if 1 is scanned
2   write 0
     go left
     go to step 2 if 1 is scanned
     write 1
     go left
     write 1
     stop

5. What does the following program do?

(Hint: begin with a tape containing four adjacent ones with the arrow over the leftmost one).

1   write 0
2   go left
     go to step 2 if 1 is scanned
     write 1
3   go right
     go to step 3 if 1 is scanned
     write 1
     go right
     go to step 1 if 1 is scanned
     stop

6. Write a program that will multiply a given whole number by three.

7. Two numbers are stored on the tape. A single space separates the two numbers. For example, if the two numbers are 4 and 3 we will have:

0	1	1	1	1	0	1	1	1	0	0	0

The leftmost one is being scanned. Write a program that will add the two numbers.

8. A number is printed on the tape and the machine is scanning the leftmost 1. Write a program that will write 1 if the number is even and 0 if the number is odd.

4. Conflicting computations

We have seen a little of the history and the theory of computation. Now let us see how computers solve three very simple problems.

1. Add 0.001 to itself 1000 times.
2. Let

$$x_0 = 2,$$

$$x_1 = \frac{3}{x_0^2 - 2.7},$$

$$x_2 = \frac{3}{x_1^2 - 2.7},$$

$$x_3 = \frac{3}{x_2^2 - 2.7},$$

and in general let

$$x_{n+1} = \frac{3}{x_n^2 - 2.7}$$

What is the value of x_{50}?

3. Find the smallest positive number e such that $1 + e \neq 1$.

We obtain solutions to these problems using the following programs, written in another computer language called BASIC.

```
1.  10    s = 0
    20    c = 0
    30    c = c + 1
    40    s = s + 0.001
    50    if c < 1000 then goto 30
    60    print c, s
    70    stop
2.  10    c = 0
    20    x = 2
    30    c = c + 1
    40    x = 3 / (x² − 2.7)
    50    print c, x
    60    if c < 50 then goto 30
    70    stop
```

```
3. 10      s = 1
   20      c = 0
   30      c = c + 1
   40      s = s / 2
   50      t = 1 + s
   60      if t > 1 then goto 30
   70      s = 2*s
   80      print c - 1, s
   90      stop
```

I ran each of these programs on:
  Research Machines PC4335 Accelerator;
  Acorn A3010;
  Texas TI82 Calculator.
The results were as in the table:

	Program 1	Program 2	Program 3
Research Machines	1000, 0.9999907	4.071654×10^{-2}	23, 1.192093×10^{-7}
Acorn	1000, 0.999999976	1.11218356	32, $2.32830644 \times 10^{-10}$
Texas TI82	1000, 1	1.300583644	44, $5.68434189 \times 10^{-14}$

I hope you are as shocked by these results as I am. Identical programs running on different makes of computer have given wildly different answers. Try these programs on your computer. Does your computer agree with any of the above answers or have you got a different set of answers again?

This is alarming. To see how it could have happened, we need to see how computers store numbers and we need to introduce *binary numbers*.

5. Binary numbers

What is meant by 307?

$$307 = 3 \times 100 + 0 \times 10 + 7 \times 1$$

or

$$307 = 3 \times 10^2 + 0 \times 10^1 + 7 \times 10^0.$$

(If 10^0 looks strange to you, remember that $10^0 = 1$. Any number raised to the power zero is one. So $2^0 = 1$ (we will use this later), $5^0 = 1$, $1.38^0 = 1$, and so on.) In the same way,

$$28.675 = 2 \times 10 + 8 \times 1 + 6 \times \frac{1}{10} + 7 \times \frac{1}{100} + 5 \times \frac{1}{1000}$$

or

$$28.675 = 2 \times 10 + 8 \times 10^0 + 6 \times \frac{1}{10} + 7 \times \frac{1}{10^2} + 5 \times \frac{1}{10^3}.$$

Observe that if 10 is acting as a *place holder*, we can use only the numbers 0 to 9 to multiply the powers of 10. We say we are operating in base 10. What if, instead of using 10 as a place holder, we used only 2? Then we would be allowed to use only the numbers 0 and 1 to multiply the powers of 2.

If we wrote 101 in base 2, we would mean

$$101 = 1 \times 2^2 + 0 \times 2^1 + 1 \times 2^0 = 5 \text{ (in base 10)} \quad \text{(Recall that } 2^0 = 1\text{.)}$$

Also,

$$101111 = 1 \times 2^5 + 0 \times 2^4 + 1 \times 2^3 + 1 \times 2^2 + 1 \times 2^1 + 1 \times 2^0 = 47 \text{ (in base 10)}.$$

We can also use a point. 11.11101 would mean

$$11.11101 = 1 \times 2 + 1 \times 2^0 + 1 \times \frac{1}{2} + 1 \times \frac{1}{2^2} + 1 \times \frac{1}{2^3} + 0 \times \frac{1}{2^4} + 1 \times \frac{1}{2^5}$$

$$= 2 + 1 + \frac{1}{2} + \frac{1}{4} + \frac{1}{8} + 0 + \frac{1}{32}$$

$$= 3 \frac{29}{32}$$

$$= 3.90625 \text{ (in base 10)}.$$

Numbers written using the base 2 are called *binary numbers*. We often use a small 2 to indicate that a number is written in base 2 (e.g. 110_2).

The importance of binary numbers is that computers store information using binary numbers. We shall see how in Section 6. But first, we need some notation. We need to think about writing numbers in *floating point form*.

Floating point form

We can write the number 423.8572 in the form 0.4238572×10^3.

When we do, 0.4238572 is called the *mantissa* and the power to which 10 is raised is the *exponent*.

Similarly, 0.0000583716 may be written in the form 0.583716×10^{-4}. Now 0.583716 is the mantissa and the exponent is -4.

The mantissa must not have a 0 as the first digit after the decimal point. If there was a zero in that place, we would simply move the decimal point along one place and reduce the exponent by one. After the decimal point, the mantissa must have a digit greater than or equal to 1. Because of this, the mantissa must always be greater than, or possibly equal to, 0.1. Also, because there is no non-zero digit to the left of the decimal point, the mantissa must be less than 1.

In general, when we write a number in the form $m \times 10^e$, where $0.1 \leq m < 1$, we say the number is in floating point form. We say that m is the mantissa and e the exponent of the floating point number.

Binary numbers may also be written in floating point form. They too have a mantissa and an exponent. We may write the number $10111.101 = 0.10111101 \times 2^5$ (the exponent here is written in decimal form). But since $5 = 101_2$ we have:

$$10111.101 = 0.10111101 \times 2^{101}.$$

The mantissa is 0.10111101 and the exponent is 101.

Note that in binary numbers, the digit to the right of the point *must* be a 1.

Why is it called 'floating point form'? I don't know. But one explanation might be that just as a cork moves up and down on the surface of the sea, but always remains on the surface, so in floating point form, the decimal point moves through the digits during various decimal operations, but always stays to the left of the first non-zero digit of the answer. Can you think of a better explanation?

Worksheet 2

1. (a) Write 11_2 in base 10.
   (b) Write 11.1_2 in base 10.
   (c) Write 11.11_2 in base 10.
2. Write the following decimal numbers in base 2.
   (a) 8 (b) 15 (c) 15.75 (d) $3\frac{5}{16}$
3. Write the following binary numbers in floating point form.
   (a) 1011.10111 (b) 0.0010111 (c) 1.1010101 (d) 11011011
   State, in each case, the mantissa and the exponent.

   We have stated that when these binary numbers are written in the form $m \times 2^e$ the first digit of m must be a 1. Hence, we must have $m \geq \frac{1}{2}$. (Why?)
4. Find the mantissa and the exponent of the following numbers.
   (a) 111.11111_2 (b) 0.000011011_2 (c) 1110101_2 (d) 3.625
5. In this question, you are allowed to use only 8 binary digits (0 or 1).
   If $x = 0.10110000$, find the two eight-binary-digit numbers that are closest to x.
   (Hint: look for the smallest number that is larger than x and look for the largest number that is smaller than x.)
   Write x and the two numbers you have found in decimal form.
6. n is a number whose mantissa has six (binary) digits and whose exponent has four (binary) digits.
   Write down the largest possible value of n.
   Write down the smallest possible value of n.
7. Multiply the binary numbers $0.11 \times 2^{10} \times 0.101 \times 2^{11}$. Express your answer in floating point form. (Observe how the point 'floats'.)

6. Limitations on storage in computers

We now discuss calculations as performed on a computer. We need to know, then, how computers store numbers and how error terms in computer arithmetic both arise and are measured.

How computers store numbers

When a number is stored in a computer's memory, it is written as a binary number and it is stored in floating point form. (Floating point form allows a far greater range of numbers to be stored in the computer's memory.) So we need to find somewhere to store the mantissa, the exponent and the signs of each of these numbers.

Different makes of computer have different memory allocations for m and e. We shall describe a system that would not be unreasonable in practice.

Suppose that in a computer's memory:
1 digit stores the sign of m,
1 digit stores the sign of e,
23 digits store the value of m,
7 digits store the value of e.

This system has 32 digits altogether, as shown in Fig. 2. But recall that the first digit of m is always 1.

Hence this digit does not need to be stored in the computer's memory. So the 23 digits available for m may be used to store the 2nd, the 3rd, ..., the 24th digits of m. This gives a 24-binary-digit mantissa stored in 23 binary digits of memory.

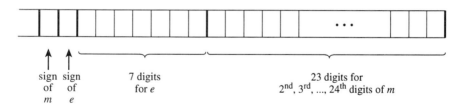

Fig. 2. Computer memory

The number $-0.110110101 \times 2^{-1011010}$ would be stored as

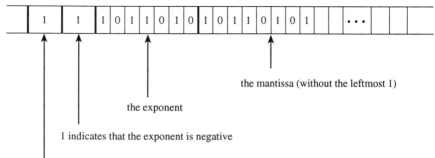

We shall call a number that can be stored in our computer's memory a *computer number*.

Errors in computer arithmetic

As a first illustration, we ask the following question: what is the smallest positive computer number e for which $1 + e > 1$?

We must have $1 + e = 1.000000000000000000000001$
$$= 0.\underbrace{100000000000000000000001}_{\text{24 binary digits}} \times 2^1$$

in the floating point form.

So
$$e = 0.000000000000000000000001$$
$$= 2^{-23}$$
$$= 1.192\,093 \times 10^{-7}$$

Does this number ring a loud bell? Look back to the end of Section 4. There you will see that this same number was the value found for e in program 3 by the Research Machines computer.

This suggests that the Research Machines computer stores numbers in a 24-binary-digit mantissa. (The 23 digits in the machine together with the first, known to be a 1.) This suggests also that the Acorn computer stores numbers in a 33-binary-digit mantissa while the Texas TI 82 uses a 45-binary-digit mantissa. And this suggests one reason why the three computers gave different answers to the three identical programs. Each of the computers has a different way of storing numbers.

Run program 3 on your computer. What is the first number printed out? What is the size of the mantissa on your computer?

What would be the smallest number that could be stored in our computer? Or, to put it another way, what is the smallest computer number for this system? We would need the mantissa to be as small as possible: so $m = 0.5$. The exponent should be negative and numerically as large as possible. So the smallest number that can be stored in our computer is:

$$0.100000\ldots \times 2^{-1111111} = 0.5 \times 2^{-127} \approx 2.939 \times 10^{-39}$$

The largest computer number for this system is

$$0.111111111111111111111111 \times 2^{1111111} \approx 1.7 \times 10^{38}$$

It may appear that with such a huge range of numbers at our disposal, there could be no problems with accuracy. But these numbers provide a false sense of security. Observe the following.

There are two choices for the 2nd digit of m.
There are two choices for the 3rd digit of m.
$$\vdots$$
There are two choices for the 24th digit of m.

So there are $2^{23} = 8\,388\,608$ different numbers in our computer available for use as a mantissa. This may sound impressive: but wait! What is going on in the exponent as we are looking at these $8\,388\,608$ numbers in the mantissa?

If exponent $e = 0$ then the numbers range from $0.1000\ldots0 \times 2^0$ to $0.1111\ldots111 \times 2^0$. This is the interval $[\frac{1}{2}, 1)$ and there are $8\,388\,608$ different numbers in our computer than can be stored with perfect accuracy in this interval.

If exponent $e = 1$, then the numbers range from $0.1000\ldots0 \times 2^1$ to $0.1111\ldots111 \times 2^1$. This is the interval $[1, 2)$ and again we can store $8\,388\,608$ numbers with perfect accuracy in this interval.

In the same way, if the exponent e, of a number is 2, then the number must lie in the interval $[2, 4)$ and once again, there are $8\,388\,608$ numbers that can be stored with perfect accuracy in this interval.

Do you see a problem lurking over there on the horizon? The intervals are doubling in length but the number of perfectly accurate computer numbers remains constant, at $8\,388\,608$.

If the exponent of a number is 24, then that number lies in the interval $[\frac{1}{2} \times 2^{24}, 1 \times 2^{24}) = [2^{23}, 2^{24})$. The length of this interval is $2^{24} - 2^{23} = 8\,388\,608$, so in this interval, only the integers (whole numbers) have a slot in the computers memory. Every number that is not an integer must be rounded.

Let's go into the heart of the problem.

If a number has exponent 51, then that number lies in the interval $[2^{50}, 2^{51})$. The length of this interval is $2^{51} - 2^{50} \approx 1\,125\,899\,907\,000\,000$. Since the computer can store $8\,388\,608$ numbers in the interval with perfect accuracy, there is a gap of approximately

$$\frac{1\,125\,899\,907\,000\,000}{8\,388\,608 - 1} = 134\,217\,744$$

between adjacent numbers stored in the computer's memory.
(Observe that there are $8\,388\,608 - 1$ gaps between the numbers.)

Think about that. If you are sitting on one of these numbers, your nearest neighbour is $134\,217\,744$ whole numbers away from you. If you set out to walk towards your neighbour and you moved at the rate of one whole number per second, never stopping, it would take you $4\frac{1}{4}$ years to reach him.

What happens if you want to use a number that lies somewhere in between you and your neighbour? The simple, but uncomfortable truth is that you can't. The computer will take the nearest computer number to the one you require (that will be either you or your neighbour) and use that. So the error could be as much as $\dfrac{314\,217\,744}{2} = 67\,108\,872$. Awesome!

We come, finally, to what every good story should possess: the moral. Be careful when you use a computer. Be especially careful when using very large (or very small) numbers. Do not believe that an answer must be correct because it has been worked out on a computer. Use your brain, your intelligence and your common sense. Then you and your computer should get on together very well.

Solutions to Worksheet 1

1. Initially, the tape contains a single 1. After the program has run, the tape contains two ones.
2. If the tape initially contains the number n, after the program has run, the tape contains $n + 2$.
3. 1   go right
   - go to step 1 if 1 is scanned
   - go left
   - write 0
   - go left
   - write 0
   - stop

   If the machine is initially scanning the leftmost 1 in the block of ones, the instructions would not change at all.
4. If the machine initially contains the number n, the program calculates $n - (n + 2)$.
5. The program doubles whatever is initially on the tape.
6. 1   write 0

   2   go left
   - go to step 2 if 1 is scanned
   - write 1
   - go left
   - write 1

   3   go right
   - go to step 3 if 1 is scanned
   - write 1
   - go right
   - go to step 1 if 1 is scanned
   - stop
7. 1   go right
   - go to step 1 if 1 is scanned
   - go right
   - go to step 3 if 0 is scanned
   - go left
   - write 1

2   go right
        go to step 2 if 1 is scanned
        go left
        write 0

3   stop
8. 1   write 0
        go right
        go to step 100 if 1 is scanned
        stop
   100   go right
            go to step 100 if 1 is scanned
            go right
            go to step 101 if 1 is scanned
            write 1
            go left
            go to step 102 if 0 is scanned
   101   write 0
            go left
   102   go left
            go to step 102 if 1 is scanned
            go right
            go to step 1 if 1 is scanned

Solutions to Worksheet 2

1. (a) 3   (b) 3.5   (c) 3.75
2. (a) 1000   (b) 1111   (c) 1111.11   (d) 11.0101
3. (a) 0.101110111, 2^{100} (b) 0.10111, 2^{-10} (c) 0.11010101, 2^1 (d) 0.11011011, 2^{1000}: $m \geq \frac{1}{2}$ because 0.1 represents $\frac{1}{2}$
4.   mantissa   exponent
   (a)   0.11111111   11
   (b)   0.11011 -100
   (c)   0.1110101   111
   (d)   0.11101   10
5. The largest eight-digit number that is smaller than x is 0.10101111 = 0.683 593 75.
   The smallest eight-digit number that is larger than x is 0.10110001 = 0.691 406 25.
6. $0.111111 \times 2^{1111} \approx 0.984\,375 \times 2^{15} - 32\,256$
   $0.100000 \times 2^{-1111} \approx 0.5 \times 2^{-15} = 0.000\,015\,3$
7. 0.1111×2^{100}

References and Further Reading

L. Casti, *Five golden rules: great theories of 20th century mathematics – and why they matter* (John Wiley, 1995)

M. Davis, 'What is a computation?', *Mathematics today, twelve informal essays* (ed. L. A. Steen), (Springer-Verlag, 1978) 241–267

K. Devlin, *Mathematics: the new golden age* (Penguin Books, 1990)

A. Hodges, *The enigma of intelligence* (Counterpoint, 1983)

R. Penrose, *The Emperor's new mind* (Oxford University Press, 1989)

J. Rotman, *Theory of groups, an introduction* (Allyn Bacon, 1965)
(This is somewhat more advanced, but it contains an excellent treatment of Turing Machines.)

3 Games of Chance

by DAVID HARRIES

Department of Science and Technology Education, University of Reading

1. Starting activities

The purpose of this chapter is to introduce you to ideas of probability through an approach involving games. By the end of the work you will

- have been introduced to basic ideas of probability and expectation;
- be in a position to extend your own concepts and ideas by further reading and exploration;
- be able to further develop your analytical problem solving skills, including planning and designing appropriate situations for experiments and games.

Some of the tasks are best carried out as part of a group. Although this may not always be possible, it is important that you discuss your ideas with others whenever you can, so that you clarify and extend your own thinking and communication skills. Individual work is also needed in analysing and designing the situations. Answers to the questions within the text are given in the discussion in Section 5. Above all else enjoy your exploration!

You can start with these three experiments which show that caution is needed in applying intuitive ideas to probability arguments! They are analysed in some detail in Section 5 but try them out first.

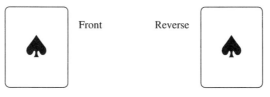

Card game

Three special cards are needed. The first card has a spade symbol in the centre of one side and another spade symbol on the reverse i.e. spade-spade as shown in the diagram above. In the same way the next card needs to be a heart-heart, and the third one a spade-heart.

Put them into a box so that they cannot be seen. Mix them up and take one out and place it on a table without looking at the underside.

Suppose a spade is uppermost. You bet with a friend (or group) that the bottom of the card is the same as the top. There are only two possibilities, spade-spade or spade-heart, so that the probability of winning will be the same for each party. Try it out several times. What happens? Why?

Think of a number
Ask a group of about 20 friends, or a class of pupils, to think of a number between 1 and 10 instantly, when you give the word. Record the results. It *must* be the first number that pops into their head.

There is an equal chance of choosing any of the numbers, so that the frequency of occurrence of each number ought to be the same.

What happens?

Birthday problem
Consider the following.

If you have 26 people in a room, the probability that there are two with the same birthdate e.g. April 5th (not necessarily the same year) will be $26/365 \times 25/365$, if we ignore leap years, which is very low! Try to find a group of people, ideally about 26 or more, who will be willing to bet (a sportsman's bet!) that there won't be two people with the same birthdate. You bet that there will be! Try the experiment with the group; you may be surprised! Whether you can carry out the experiment or not, try to analyse the problem in more detail.

These three examples show the caution needed when thinking about probability. We need to consider a formal definition of probability, along with basic ideas of combining probabilities.

2. Probability

There are two approaches in defining probability that we can use.

Theoretical

If an outcome set has n equally-likely outcomes, and an event set can occur in a ways, then the probability of the event happening is a/n.

E.g. when a dice is thrown the probability of getting a six, $p(6)$, is $1/6$.

Using this definition we can calculate some probabilities without carrying out an experiment.

Experimental

If a large number of trials, say n, are carried out and an event occurs a times then an *estimate* of the probability of the event occurring is a/n.

E.g. we could throw a dice 100 times and count the occurrences of each number to see if it is a fair dice. If a 6 occurred 75 times, say, then experimentally $p(6) = 75/100 = 3/4$; perhaps not a fair dice!

This definition allows us to estimate a probability after carrying out an experiment. Some situations can only be analysed in this way, e.g. which horse will win a race!

Question 1
A normal pack of 52 cards is shuffled and placed face down. The cards are turned over one by one.

Find the probability that:

(a) the first card is a Spade;
(b) the first card is Ace of Spades;
(c) the last card is the King of Hearts;
(d) the second card is not a Jack if the first card was 2 of Spades.

The answers are in Section 5.

Laws of probability

Events are *independent* if the outcome of one does not affect the other, e.g. the probability of getting a tail during the second toss of a coin is not affected by the result of the first toss.

If two events A and B are independent then the probability of *both* occurring

$$p(A \ and \ B) = p(A) \times p(B).$$

Coin experiment
During an experiment a coin is tossed twice; what is the probability of getting two heads (H) in a row?

$$p(H) = \tfrac{1}{2}.$$

We can write $H + H$ as heads followed by heads.

Each throw is independent, therefore

$$p(H + H) = \tfrac{1}{2} \times \tfrac{1}{2} = \tfrac{1}{4}$$

We can also illustrate this by listing all the possible outcomes for coin 1 and coin 2, where H is heads and T is tails.

$$H + H \qquad H + T \qquad T + H \qquad T + T$$

so that $p(H + H) = \tfrac{1}{4}$.

You can explore the following situations yourself, or consider your own questions.

What about three tosses and three heads?

What is the probability of getting ten heads in a row?

Question 2
From the situation in Question 1, what would be the probability of the first four cards being Aces?

The solution is given in Section 5.

For two independent events A and B the probability of *either* occurring i.e. the probability of A occurring or B occurring or both A and B occurring, is given by

$$p(A \ or \ B) = p(A) + p(B) - p(A \ and \ B)$$

For example, when drawing a card from a pack the probability of getting a Heart or a picture card is:

$$p(\text{heart}) + p(\text{picture}) - p(\text{heart and picture}) = \tfrac{13}{52} + \tfrac{12}{52} - \tfrac{3}{52}$$

$$= \tfrac{22}{52}$$

$$= \tfrac{11}{26}$$

Can you explain why it's not just $p(\text{heart}) + p(\text{picture})$?

If you have used Venn diagrams before you may find them useful. An explanation is given in Section 5.

Tree diagrams

Probabilities of two or more events can be illustrated on a *tree diagram* , showing all the possible outcomes.

E.g. a dice is thrown twice. Find the probability of throwing an even number less than 6 on both occasions.

'An even number less than 6' means throwing a 2 or 4, which is written as 2,4 on the tree diagram.

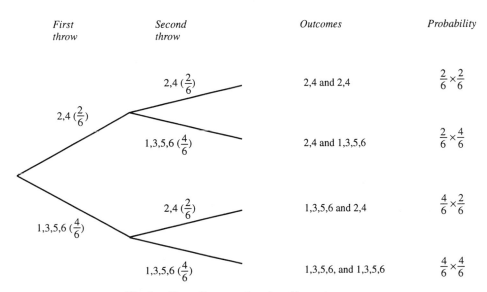

Fig. 1. Tree diagram showing dice outcomes

We can see that the probability of throwing an even number less than 6 on both occasions is

$$p(2 \text{ or } 4 \text{ } and \text{ } 2 \text{ or } 4) = \tfrac{2}{6} \times \tfrac{2}{6} = \tfrac{1}{9}$$

Question 3
A game is played by drawing a card from an ordinary pack, and throwing a dice.

A player wins the game by drawing a heart from the pack, and throwing a 2 or 4 on the dice. Use a tree diagram to show all the outcomes, and find the probability that she wins.

The solution is given in Section 5.

Worksheet 1

1. It will either rain tomorrow or it will not, therefore the probability of rain tomorrow is $\tfrac{1}{2}$. Do you agree or not? Why?
2. A bag contains 3 red balls, 5 black balls and 2 white balls. What is the probability that a ball drawn at random is
   (a) red;
   (b) not red;
   (c) either red or black;
   (d) neither red or black;
   (e) red, black or white?
3. Four brown and three green leaves are placed in a bag. Two leaves are drawn at random. Use a tree diagram to find the probability that
   (a) the first leaf is brown and the second is green;
   (b) both leaves selected are green;
   (c) neither leaf is purple;
   (d) one green and one brown leaf is obtained.
4. Out of 28 people surveyed in a street 18 own a red car and 9 own a blue car. Two did not own cars. If one of the group was chosen at random what would be the probability that he or she
   (a) does not own a car;
   (b) owns a blue car;
   (c) owns a red or blue car;
   (d) owns more than 1 car?
5. (a) 'Honest Harry' offers players on a street corner the following game. He rolls a red dice, and the player rolls a blue dice. If the combined score is 8 or more the player wins. Would you advise someone to play Harry's game?
   (b) Harry decides to change the rules: if the total score is 8 or more the player wins outright. If the total is between 3 and 7 inclusive they roll again, and if this new roll is a pair the player wins. Harry maintains that he can only win outright with one total, i.e., 2 which is not likely, so he's giving away money! Would you play the game with Harry!
   (c) What happens if Harry wins outright on a score of 2 or 7?
   (d) Devise your own version of Harry's game!

3. Expectation, arithmetic mean and standard deviation

We now have a means of calculating the probability of events, either from analysis of all the outcomes, or obtaining an estimate from an experiment. However, in a game it isn't only the probability that's important but how much it costs to play and how much we can win.

There are some definitions we will need for later work.

A variable can be *discrete* or *continuous*. A *discrete* variable can only take certain values while a *continuous* variable can take any value. This is best shown by an example. Shoe size is a discrete variable, since you can get a size 7 or 8 but not 7.34 or 7.2 say. Height, however, is continuous, since it can take any value such as 7.8565 metres or 7.29 metres.

The *arithmetic mean* average, or *mean*, is a number which can be used to represent a set of data. It is calculated by finding the total of the data values and then dividing by the number of data values which occur. For example, the *mean* of 5, 12 and 13 is 30 divided by 3, which gives 10.

Now we can use these ideas to explore some situations further.

Consider the following game.

Anita suggests to Barbara that they play a dice game. Anita throws the dice. If a 6 occurs she will pay £6 to Barbara, if a 1 occurs she will pay £1 to Barbara. If any other number occurs she wins £2 from Barbara. Should Barbara play?

$$p(\text{Anita throwing } 6) = \tfrac{1}{6}$$

$$p(\text{Anita throwing } 1) = \tfrac{1}{6}$$

If Anita throws a 6 Barbara will win £6.

Therefore, Barbara is likely to win £6 $\tfrac{1}{6}$ of the time.

Similarly, Barbara is likely to win £1 $\tfrac{1}{6}$ of the time.

Also Barbara is likely to lose £2 $\tfrac{4}{6}$ of the time.

In 600 goes Barbara can therefore expect to win

$$(100 \times £6) + (100 \times £1) - (400 \times £2) = -£100.$$

Dividing by 600 gives a mean average expected win per game of

$$(\tfrac{1}{6} \times £6) + (\tfrac{1}{6} \times £1) - (\tfrac{4}{6} \times £2) = -£\tfrac{1}{6}$$

Therefore Barbara's expected winnings per game are $-£\tfrac{1}{6}$ and she should not play!

This quantity, the mean average expected win per game for Barbara, is defined as Barbara's *expectation*. It can also be found by multiplying the amount won in each case by the probability of that case occurring i.e.

$$(£6 \times \tfrac{1}{6}) + (£1 \times \tfrac{1}{6}) - (£2 \times \tfrac{4}{6}) = -£\tfrac{1}{6}$$

We can now state a more formal definition of *expectation* for a random variable. (A random variable is a quantity which occurs as the result of carrying out an *experiment*. In the game above the amount won by Barbara is a random variable.)

The *Expectation* of a random variable R which has values $r_1, r_2, r_3, \ldots, r_n$ where n is a positive integer, each with probability $p_1, p_2, p_3, \ldots, p_n$ respectively, is

$$E(R) = p_1 r_1 + p_2 r_2 + p_3 r_3 + \cdots + p_n r_n$$
$$= \sum_i p_i r_i$$

E.g. a box contains 3 blue, 2 white and 4 green balls. Alan chooses a ball. If it's green he wins £1 from Joe, if it's white no one wins and if it's blue Alan pays Joe £2. What is Alan's expectation?

$$p(\text{blue}) = \tfrac{3}{9}, \qquad p(\text{white}) = \tfrac{2}{9}, \qquad p(\text{green}) = \tfrac{4}{9}$$

Alan's expectation is given by

$$(\tfrac{3}{9} \times (-£2)) + (\tfrac{2}{9} \times £0) + (\tfrac{4}{9} \times £1) = -£\tfrac{2}{9}$$

i.e. Alan should not play!

What is Joe's expectation?

Question 4
In a game a player throws two dice. If the total score of the dice is less than 7 she wins that total amount shown on the dice in £. If the total is greater than or equal to 7 she has to pay £10 to the other player. What is her expectation?

Should she play the game?

In a new version of the game she pays £10 to throw the dice, and wins the 'total' in £ if it is less than 7 or greater than 10. Should she play now?

Develop this idea for a simple game where the expectation for the player is positive!

The solution is given in Section 5.

For a discrete random variable R, such as we are concerned with, the mean is the same as the expectation. We can represent the mean by the Greek letter mu (μ), so that

$$\mu = E(R),$$
$$\mu = \sum_i p_i r_i.$$

Game 1
A fairground game consists of throwing an unbiased four-sided dice with scores of 18, 20, 22, and 32 marked on respective faces. A player wins by scoring more than 100 with four consecutive throws. Ben plays the game and scores 22, 18, 32 and 20 respectively. His mean score is $\tfrac{92}{4} = 23$.

Since the dice is fair the probability of each score occurring is $\frac{1}{4}$. We can see that the mean and the expectation are the same in this case.

In practice, if Ben threw the dice many times, we would expect 18 to occur for approximately one quarter of the throws, 20 for approximately one quarter of the throws, etc. The mean for all the throws would therefore be approximately the same as the expectation, the greater the number of throws the greater the agreement between the two.

Game 2

Suppose now that the dice is in fact biased so that the probability distribution of the scores is as follows:

Score	Probability
18	$\frac{3}{8}$
20	$\frac{1}{8}$
22	$\frac{2}{8}$
32	$\frac{2}{8}$

This means that in 8 games we would expect the scores to be:

$$18 \quad 18 \quad 18 \quad 20 \quad 22 \quad 22 \quad 32 \quad 32$$

If you work out the mean and the expectation you will see that they are the same.

Sometimes, it's also useful to know the spread of the data about the mean, e.g. the following two sets of data have the same mean of 400, however the spread is very different:

Set 1:	298	400	502		
Set 2:	1	1	2	36	1960

(In both sets each data value has an equal probability of occurring: for Set 1, p(298 occurring) $= \frac{1}{3}$, for Set 2, p(1 occurring) $= \frac{1}{5}$.)

The data in Set 1 is bunched more closely around the mean. If we wanted to know the amount of spread more precisely we could simply add all the deviations from the mean, i.e. $(298 - 400) + (400 - 400) + (502 - 400)$. What do we get as the total? What total do you get if you add all the deviations for Set 2? Is this true for all data? Can you think why this is so?

Since adding all the deviations leads to negative and positive values cancelling each other out, we need some means of making all the deviations positive. We can do this by squaring all the deviations, finding the mean of these squared values and then taking the square root.

For example, for the values 4, 10 and 19 this will give

$$\sqrt{\frac{(4-11)^2 + (10-11)^2 + (19-11)^2}{3}} = 6.164$$

This is known as the *standard deviation* of the data. The larger the standard deviation the wider the spread.

The standard deviation can be formally defined as

$$\sigma = \sqrt{E\left[(r - \mu)^2\right]} \quad \text{(where } \sigma \text{ is the Greek letter sigma)}.$$

In practice it's easier to use the following formula to carry out the calculations,

$$\sigma = \sqrt{\Sigma r^2 p(r) - \mu^2}$$

E.g. for the data in Set 1 above,

$$\sigma = \sqrt{\frac{(-102)^2 + 0 + 102^2}{3}}$$

$$\sigma = 83.28$$

or

$$\sigma = \sqrt{(298^2 + 400^2 + 502^2) \times \tfrac{1}{3} - 400^2}$$

$$= 83.28$$

The standard deviation for the second set of data will be much larger. You can work it out as an exercise!

Show that the standard deviation for the set of scores in the fairground dice examples above are 5.385 for Game 1, and 5.562 for Game 2.

As well as finding the expectation for a player in a game we can now find a statistic showing the degree of spread of the results for several games.

Consider this example.

Laura is asked by Peter to play the following game. She throws a 10-sided dice. If she throws a number less than 3 she wins £4, and wins £7 if she throws a 10. If it lands on any other number she pays £2 to Peter. Find her expectation and the standard deviation for several throws. Would you advise her to play the game?

$$p(\text{winning £4}) = \tfrac{2}{10}$$

$$p(\text{winning £7}) = \tfrac{1}{10}$$

$$p(\text{losing £2}) = \tfrac{7}{10}$$

Expected profit $= (\tfrac{2}{10} \times 4) + (\tfrac{1}{10} \times 7) - (\tfrac{7}{10} \times 2)$
$= £0.10$

The mean expected profit is only £0.10, so that she would expect to win in the long run, but only a relatively small amount.

Using $\sigma = \sqrt{\Sigma r^2 p(r) - \mu^2}$ we can find the standard deviation

$$\sigma = \sqrt{\left(4^2 \times \tfrac{2}{10} + 7^2 \times \tfrac{1}{10} + (-2)^2 \times \tfrac{7}{10}\right) - \left(\tfrac{1}{10}\right)^2}$$

$$= £3.30$$

This tells us that the spread of results is relatively large compared to the mean. Laura can expect to make a small overall profit after playing many games, but any one game result can fluctuate by a relatively large amount.

Worksheet 2

Try the following game:

A and B toss a coin each

- if they are both heads or both tails A wins £1,
- if they are different B wins £1.

1. Who do you think has the best chance of winning?
   Work out the expectations.
2. The game is changed so that A tosses 2 coins and B tosses 1 coin:
   - if they all match A wins £1 (i.e 3 heads or 3 tails)
   - if they are different B wins £1
     (a) How would you advise A or B about playing?
         How could you make these equal?
     (b) Analyse the situation for the rule:
   - if there are at least 2 tails A wins £1 otherwise B wins £2.
3. Modify the rules to the coins game to make it

   - fair
   - unfair on A

   Can you find a way of making it apparently fair, but in fact biased in A's favour?
4. Develop a simple dice or card game similar to the earlier example involving Laura. You do not need to restrict your game to a six-sided dice, but keep it simple to begin with. Calculate the mean profit and standard deviation for your game. Alter your game slightly, and try to predict and analyse the effects on the mean and standard deviation.
5. Design a game using any suitable equipment, e.g. dice, coins, squared paper, cards, which can be used in a fairground. The game must be appealing, biased (only slightly!) in favour of the 'operator', but not so much as to put off people from spending money! Try it on your friends! You can develop the game by analysing the maths involved first, or develop your ideas by trial and improvement. Whichever, be sure to work out the probabilities and expectations.

4. Final thoughts

You have now had some opportunities to consider ideas of probability and expectation. As you have seen, the process of making decisions is more complex than we might think, and despite logical analysis is often linked to our emotions, hence the reason why so many people enter the National Lottery, despite a poor chance of winning, approximately 1 in 14 million of getting the jackpot!

Now that you have started to look more closely at probability and games you may like to continue further. There are ideas such as conditional probability, games theory and statistical analysis that are worth pursuing. A brief list of books for possible investigation is given at the end of this chapter. Hopefully you will have a lot of fun in exploring these ideas further!

5. Discussion of experiments and questions

Card game
The card game probabilities are $\frac{2}{3}$ for both sides being the same and $\frac{1}{3}$ for them being different. This is because there are two chances of the card in our example being spade-spade. If you label the sides spade1 and spade2, then it could be spade1−spade2 or spade2−spade1. And there is only one chance for spade-heart! (This game is based on the three-card swindle discussed in Martin Gardner's book *Aha! Gotcha* referred to at the end of this chapter. It contains many examples of intriguing ideas and paradoxes which are well worth reading.)

Think of a number
When we choose numbers they are not random! Most people will choose 7, or a number near it so that the distribution is uneven. Why 7 you ask? Who knows!

Birthday problem
This is also more complex than the given argument would indicate. The correct analysis is given below. If you have 50 people or more in a room your chances of winning would be *very* high!

Ignore leap years, suppose the 1st person's birthday is on a particular date.

$$p(\text{2nd person's birthday is different}) = \frac{364}{365}$$

$$p(\text{3rd person's birthday is different}) = \frac{363}{365}$$

$$p(\text{4th person's birthday is different}) = \frac{362}{365}$$

$$p(n\text{th person's birthday is different}) = \frac{365 - (n - 1)}{365}$$

Therefore,

$$p(\text{all birthdays are different}) = \frac{365}{365} \times \frac{364}{365} \times \frac{363}{365} \times \cdots \times \frac{365 - (n-1)}{365}$$

$$= \frac{365 \times 364 \times 363 \times \cdots \times (365 - (n-1))}{365^n}$$

Now, all the birthdays for the group are different, or two or more people have the same birthday.

Therefore

$$p(\text{all birthdays are different}) + p(\text{at least 2 birthdays are the same}) = 1$$

so that

$$p(\text{at least 2 birthdays are same}) = 1 - p(\text{all birthdays are different})$$

$$= 1 - \frac{365 \times 364 \times 363 \times \cdots \times (365 - (n-1))}{365^n}$$

For $n = 23$, $p = 0.5$
For $n = 50$, $p = 0.97$ which is surprisingly high.

(There is a good discussion of many interesting points involving probability, such as the birthday problem, in Warren Weaver's book listed at the end of the chapter.)

Question 1
(a) $\frac{13}{52}$ or $\frac{1}{4}$ (b) $\frac{1}{52}$ (c) $\frac{1}{52}$ (d) $\frac{47}{51}$

Coin experiment
Probability of 3 heads in 3 tosses of a coin is $\frac{1}{2} \times \frac{1}{2} \times \frac{1}{2} = \frac{1}{8}$.
Probability of 10 heads in a row is $(\frac{1}{2})^{10}$.

Question 2
The probability of the first four cards being Aces is $\frac{4}{52} \times \frac{3}{51} \times \frac{2}{50} \times \frac{1}{49} = 0.000\,003\,7$.

Explanation of $p(\text{A or B}) = p(\text{A}) + p(\text{B}) - p(\text{A and B})$

If A and B cannot both occur then $p(\text{A or B})$ is the same as $p(\text{A}) + p(\text{B})$. In this case A and B are said to be *mutually exclusive*.

For example, if A = {throwing a 1 on a dice} and B = {throwing a 6 on a dice} then $p(\text{A or B}) = \frac{1}{6} + \frac{1}{6} = \frac{2}{6}$.

However, if A and B have something in common, as in the example given, then by simply adding $p(\text{A})$ and $p(\text{B})$ we will include the common section twice. We can show this on a Venn diagram, where shading A and then shading B shows the overlap in the section A *and* B.

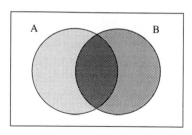

Question 3

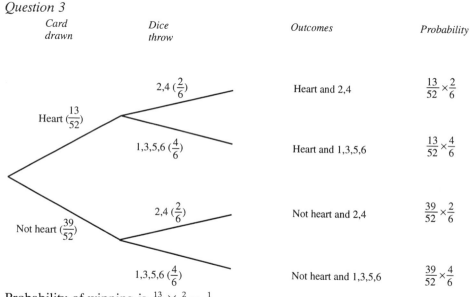

Probability of winning is $\frac{13}{52} \times \frac{2}{6} = \frac{1}{12}$

Question 4

The expectation is $-£\frac{140}{36}$. No!

The expectation is now $-£\frac{76}{36}$. It's advisable not to play!

Solutions to Worksheet 1

1. It may or may not rain, but there is no reason why each should be equally likely!
2. There are 10 balls altogether:
   (a) $p(\text{red}) = \frac{3}{10}$
   (b) $p(\text{not red}) = \frac{7}{10}$
   (c) $p(\text{either red or black}) = \frac{8}{10} = \frac{4}{5}$
   (d) $p(\text{neither red nor white}) = \frac{5}{10} = \frac{1}{2}$
   (e) $p(\text{red, black or white}) = \frac{10}{10} = 1$

3.

| First leaf | Second leaf | Outcome | Probability |

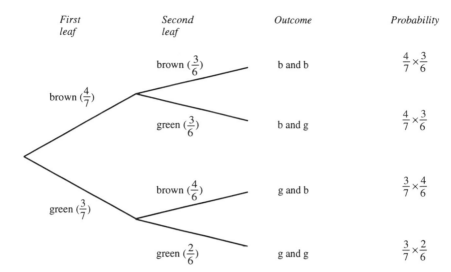

(a) $\frac{2}{7}$ (b) $\frac{1}{7}$ (c) 1

(d) We can either have a brown leaf followed by a green leaf, or a green leaf followed by a brown leaf, so that the probability is given by $(\frac{4}{7} \times \frac{3}{6}) + (\frac{3}{7} \times \frac{4}{6}) = \frac{4}{7}$

4. $28 - 2 = 26$ own cars

We know that 18 own a red car and 9 own a blue car, however, this gives a total of 27. Since only 26 own cars we can only conclude that one owns two cars. We can show this on a Venn diagram:

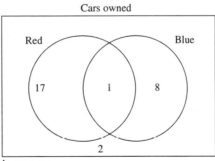

Therefore the solutions are:

(a) $\frac{2}{28} = \frac{1}{14}$ (b) $\frac{9}{28}$ (c) $\frac{26}{28} = \frac{13}{14}$ (d) $\frac{1}{28}$

5. (a) If we use the notation (red dice, blue dice) there are 15 combinations for scoring 8 or more:

$$
\begin{array}{ccccc}
(6,6) & (6,5) & (6,4) & (6,3) & (6,2) \\
(5,6) & (5,5) & (5,4) & (5,3) & \\
(4,6) & (4,5) & (4,4) & & \\
(3,6) & (3,5) & & & \\
(2,6) & & & & \\
\end{array}
$$

Therefore, $p(\text{win for player}) = \frac{14}{36} < \frac{1}{2}$

No, don't play!

(b) $p(\text{player winning}) = p(8 \text{ or more}) + p(3 \text{ to } 7 \text{ inclusive}) \times p(\text{a pair})$

$$= \frac{15}{36} + \left(\frac{20}{36} \times \frac{6}{36}\right)$$
$$= \frac{110}{216}$$
$$> \frac{1}{2}$$

So it's worth playing – just!!

(c) As for (b), the tree diagram is the same but the probabilities are now slightly different.

$p(\text{player winning}) = \frac{15}{36} + \left(\frac{14}{36} \times \frac{6}{36}\right)$
$$= \frac{104}{216}$$
$$< \frac{1}{2}$$

So it's not worth playing (just)!

(d) Up to you! For example, you could vary the winning totals, the number of throws in the game or try other ideas such as combining dice rolls with coin tosses.

Solutions to Worksheet 2

1. Possibilities are:

$$\text{H} + \text{H} \qquad \text{H} + \text{T} \qquad \text{T} + \text{H} \qquad \text{T} + \text{T}$$

A and B are both equally likely to win.

$\text{E}(A) = \left(\frac{1}{4} + \frac{1}{4}\right) \times £1 - \left(\frac{1}{4} + \frac{1}{4}\right) \times £1.$
$\quad\quad = 0$

2. (a) $p(\text{A winning}) = \frac{1}{8} + \frac{1}{8}$
$$\quad\quad\quad\quad\quad\quad = \frac{1}{4}$$

$p(\text{B winning}) \quad = \frac{3}{4}$

$\text{E}(A) \quad\quad\quad = £\frac{1}{4}$

$\text{E}(B) \quad\quad\quad = £\frac{3}{4}$

B has an advantage.

The game could be made equal by B winning £1 and A winning £3, or by modifying the winning combinations.

(b) $p(\text{A winning}) = \frac{4}{8}$

$\quad \text{E}(A) \quad\quad\quad = £\frac{1}{2}$

$\quad \text{E}(B) \quad\quad\quad = \frac{1}{2} \times £2 = £1$

It's not a good idea for A to play!

3. Your own individual approach. The winning outcomes or the stakes can be modified appropriately.

(Explore the variations such as 4 coins, 3 throws, ...)

Further reading

There are many books available on games and mathematics. The following provide a starting point.

J. D. Beasley, *The mathematics of games* (Oxford University Press, 1989)

R. Bell and M. Cornelius, *Board games around the world* (Cambridge University Press, 1988)

M. Gardner, *Aha! Gotcha. Paradoxes to puzzle and delight* (W. H. Freeman and Company, 1982)

M. Gardner, *Mathematical magic show* (Penguin Books, 1985)

W. Weaver, *Lady Luck. The theory of probability* (Heinemann Educational, 1964)

4 Weather

by BRIAN HOSKINS

Department of Meteorology
University of Reading

1. What is Weather?

When we talk about the weather, we are referring to wind, rain, sunshine, frost and so on. The weatherman or woman on the television might stand in front of a chart like that in Fig. 1(a) and forecast what the weather will be tomorrow.

The weather map in Fig. 1(a) shows lines of equal sea-level pressure drawn every 8 mb from 960 mb to 1024 mb (the unit mb will be defined later). L and H mark low- and high-pressure centres. Weather fronts are marked in bold with rounded bumps for the warm front and jagged bumps for the cold front. The little circles with symbols around them show the local weather. In particular the flags show the direction of the wind and the number of tail feathers indicates the strength. The dots show the rainfall with three dots being very heavy. The satellite image is a picture of the heat being given off rather than the light reflected. High cloud is cold and is shown as white. The ocean surface is warmer than the land surface at this time of year and so it shows as darker. The deep cloud swirls around the intense low pressure system. This storm gave intense winds and rainfall over the UK and has become known as the Burns Day storm.

When we are going to apply our mathematics to understanding weather and interpreting weather charts we need to think about what exactly we mean by, for example, the wind.

2. The Earth and its atmosphere

Surrounding us we have an atmosphere, air, which is made up of a number of gases. The largest proportion, about 70%, is nitrogen. The next largest is oxygen, comprising about 28%. The atmosphere above us is held down by the gravitational attraction of the Earth, a force per unit mass of magnitude $g = 9.8\,\mathrm{ms}^{-2}$. On every square metre of the Earth's surface there is about 10 tonnes of atmosphere pressing down. This means that there is about 200 kg of air on your head! (See question 1 on Worksheet 1).

When air moves, this is what we refer to as wind. A gentle breeze is air moving at a few ms^{-1} ($10\,\mathrm{ms}^{-1}$ is about 22 miles per hour). Gale force is a mean wind of $19\,\mathrm{ms}^{-1}$ and hurricane force is $33\,\mathrm{ms}^{-1}$.

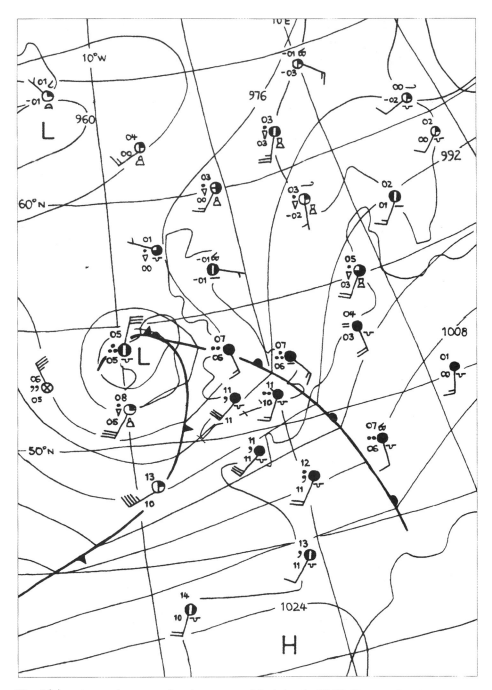

Fig. 1(a). A weather map for the eastern N. Atlantic/N.W. Europe for 6 am on 25 January 1990

Fig. 1(b). A satellite image for the same area as Fig. 1(a) at 8 am (courtesy of the Universities of Reading and Dundee)

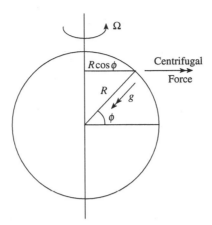

Fig. 2.   A section through the centre of the Earth

When we say that the air moves, we really mean that the air moves relative to you and me standing still on the Earth. However we must remember that we live on a planet that is spinning in space. The rotation rate of the Earth is $\Omega = 7.292 \times 10^{-5}\,\text{s}^{-1}$ and it turns $360° \equiv 2\pi$ radians in a time $2\pi/\Omega$ which is about 4 minutes less than 1 day. (See question 2 on Worksheet 1 to help you understand why it is not exactly 1 day!). Our planet, Earth, is almost spherical and has a radius, R, of about $6370\,\text{km} = 6.37 \times 10^6\,\text{m}$. Referring to Fig. 2, if we are at latitude ϕ, our distance from the axis of rotation is $R\cos\phi$, and the speed we move about this axis is $R\Omega\cos\phi$. ($\cos\phi$ is short for cosine ϕ. It is a function that you will find on your calculator. It decreases smoothly from 1 to 0 as ϕ increases from $0°$ to $90°$.) On the Equator the speed is just $R\Omega$. If you use your calculator you will find out that this speed is about $465\,\text{ms}^{-1}$. This is much faster than the speed of sound! It is also huge compared with the wind. Therefore the atmosphere almost rotates with the Earth. The small departure from this is what we refer to as wind.

Since the Earth rotates there is an apparent centrifugal force away from its axis. This is the apparent force you feel trying to throw you off if you sit on a roundabout that is rotating rapidly. Because of this centrifugal force the Earth bulges at the Equator. The distance from the centre of the Earth to the equatorial surface is about 20 km greater than that to the poles. 20 km would be a very high mountain – twice the height of Mount Everest! However, compared with the radius, R, it is very small, and we can usually treat the Earth as a sphere and absorb the centrifugal force into what we call gravity.

The final basic fact about the Earth that we need to note is that, even though the atmosphere weighs a considerable amount, it is a very thin envelope around the planet: relatively thinner than the skin of an apple. Half the mass of the atmosphere is contained within 6 km of the Earth's surface. Consequently the distance of the atmosphere from the centre of the planet can usually be taken to be R.

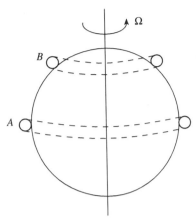

Fig. 3. A ring of air moves from the equator, position *A*, northwards to position *B*

3. The Coriolis force

We have considered gravitational attraction and centrifugal force, but there is one other crucial force that the air feels. To understand this, look at the ring of air around the equator (*A*) shown in Fig. 3. Suppose that this ring of air is initially rotating with the Earth. If it moves towards the pole to position *B*, conservation of angular momentum (the whirlpool effect) says that it must start to rotate faster than the Earth, i.e. we see an acceleration from west to east. Thus the 'southerly' wind leads to a 'westerly' acceleration. (Beware: meterologists identify winds with where they come from; oceangraphers identify currents with where they are going!).

This acceleration is associated with another apparent force, the 'Coriolis' force. As shown in Fig. 4, it always acts perpendicularly to the right of the wind direction in the northern hemisphere. Its magnitude is $2\Omega(\sin \phi)V$, where V is the wind speed. ($\sin \phi$ is short for sine ϕ. Again, you will find it on your calculator. It increases smoothly from 0 to 1 as ϕ increases from 0° to 90°.)

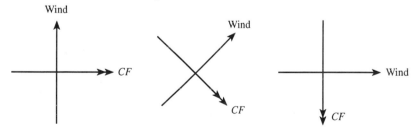

Fig. 4. The direction of the Coriolis force (*CF*) for various wind directions shown by the single arrow head

Worksheet 1

1. Work out the approximate area of the top of your head in square metres.

Remembering that the air above 1 square metre weighs about 10 tonnes and that 1 tonne $= 10^4$ kg, work out how much weight of air is pressing down on your head.

2. The Earth rotates on its axis in the same direction as it goes around the Sun. How long would a day be if the Earth turned around once on its axis in one year? To answer this, get a friend to be the Sun. Stand opposite the 'Sun' and move around the Sun, rotating once as you go around. Show that your front is always facing the Sun and that your back is always facing away from the Sun. There would be permanent daylight on one side of the Earth and permanent night on the other! The Earth actually rotates on its axis about $366\frac{1}{4}$ times in a year. Therefore there are $366\frac{1}{4} - 1 = 365\frac{1}{4}$ days in a year.

Work out the time taken for the Earth to spin on its axis, $2\pi/\Omega$, in seconds, and compare it with the number of seconds in a day (24 hours, each of 3600 seconds).

3. Work out the speed $R\Omega \cos\phi$ that you are travelling in space because the Earth is rotating. (You will need to put your latitude into your calculator and press the cos button to get $\cos\phi$).
4. The westerly wind generated by moving a ring of air from the Equator to latitude ϕ is $R\Omega(\sin\phi)^2/\cos\phi$. What would be its speed if it reached your latitude? (This time you will need to use the sin button on your calculator).
5. What is the magnitude and direction of the Coriolis force for a $20\,\mathrm{ms}^{-1}$ westerly wind at your latitude?

4. The balance of forces

We have seen that in the northern hemisphere the Coriolis force always acts to the right of the wind direction. If no other forces act, this means that the air will move around in circles. The direction will be clockwise in the northern hemisphere. Circular motions like this are sometimes found in the atmosphere (and quite often in the oceans). However they are not what we see on our weather maps.

The other horizontal force that can act on a frictionless atmosphere is the pressure force. The atmosphere feels a force driving it from high to low pressure as in Fig. 5a.

Fig. 5. (a) The pressure force (PF) from low to high pressure. (b) Geostrophic balance between the pressure force and the Coriolis force

The magnitude of this pressure force is $(p_H - p_L)/\rho d$ where ρ is the density of the atmosphere. Atmospheric pressure is usually given in millibars, mb for short, but for any calculation we must use the standard unit, the Pascal: $100\,Pa = 1\,mb$. The density of the atmosphere is conveniently close to $1\,kg\,m^{-3}$. At the surface it is about $1.2\,kg\,m^{-3}$.

We now have two horizontal forces acting on the atmosphere, the Coriolis force and the pressure force, and there is the possibility of balance between them. Such a situation is shown in Fig. 5(b). The Coriolis force acting to the right of the wind balances the pressure force which must be to the left of the wind. This is called *geostrophic balance*. For such balance the wind is along lines of equal pressure, with low pressure on the left in the northern hemisphere. The force balance is

$$2\Omega(\sin\phi)V = (p_H - p_L)/\rho d \qquad (1)$$

The wind and pressure fields are nearly in such a balance, except very close to the Equator. There, since ϕ is small, $\sin\phi$ is small (remember that both ϕ and $\sin\phi$ are zero at the Equator). Thus, from equation (1) the difference between p_H and p_L must be small. This shows why the pressure field is generally much more uniform in the tropics. Only in tropical cyclones is the wind strong enough to be associated with a strong feature in the pressure field.

5. Motion around pressure centres

Let us now think about the region around a low pressure centre as in Fig. 6(a). The pressure force is clearly inwards. Geostrophic balance is possible if the Coriolis force is outwards. Since in the northern hemisphere the Coriolis force is to the right of the wind direction, the wind must circle in an anticlockwise direction about the low pressure centre. This is referred to as cyclonic. The wind

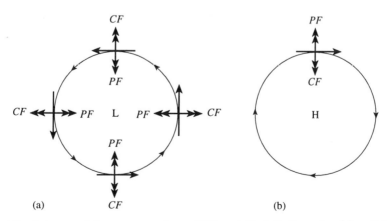

Fig. 6. The balance of the pressure force (PF) and Coriolis force for (a) anticlockwise air motion around a low pressure centre; (b) clockwise air motion around a high pressure centre

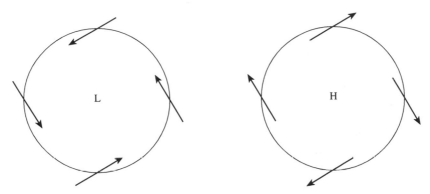

Fig. 7. The motion around low and high pressure centres when surface friction reduces the magnitude of the Coriolis force

speed is given by equation (1). Similarly, as in Fig. 6(b), there must be clockwise, anticyclonic flow around a high pressure centre. These situations occur all the time in our weather.

Near the surface of the Earth, a third horizontal force comes in to play. The rough surface provides a drag on the atmosphere. This reduces the magnitude of the wind. As a consequence the Coriolis force is not large enough to balance the pressure force which then wins the battle. The wind is reduced and turned by about 30° to blow slightly towards the low pressure and away from the high pressure (Fig. 7).

Worksheet 2

1. What is the magnitude and direction of the pressure force associated with a surface pressure decrease of 15 mb from 200 km south of you to 200 km north of you?
2. What magnitude of pressure decrease in 400 km would balance the westerly wind of $20\,\text{ms}^{-1}$ discussed in question 5 on Worksheet 1?
3. What is the average wind circling around a low pressure system with a minimum pressure near you of 981 mb, when 400 km from the centre the pressure is 1003 mb?
4. Draw a picture showing the direction of flow and of the Coriolis force and pressure force around a low pressure system in the southern hemisphere. (In the southern hemisphere the Coriolis force is to the left of the wind direction).

6. Some ideas on climate

As we are all aware the tropics are warmer than the higher latitudes because they receive more sunshine. This has two consequences that are important to us here.

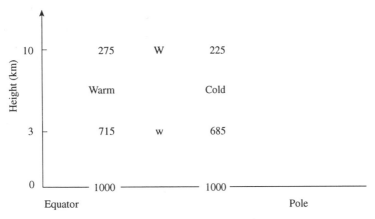

Fig. 8. Typical pressure in mb at 3 and 10 km in warm air near the Equator and cold air near the pole, assuming a uniform 1000 mb at the surface. At levels above the surface there is relatively higher pressure in the warm air and lower pressure in the cold air. This pressure difference increases with height. The corresponding pressure force is balanced by the Coriolis force associated with a westerly wind (W) which must also increase with height. Using typical values gives the average westerly wind at 10 km to be about 20 ms^{-1}. The typical maxima in the 'jet stream' are more than double this value

The pressure always increases as you move down in the atmosphere. However, in cold dense air the pressure increases more rapidly with depth than in warm air. To put it the other way around, the pressure falls more rapidly with height in the cold air. An example of this is given in Fig. 8. As seen in this Figure, if the surface pressure is almost uniform, this means that as we move away from the surface we must see relatively lower pressure in higher latitudes and higher pressure in lower latitudes. As we have seen already, there will be a wind field in balance with this pressure field. It will be a westerly wind and will increase with height (see Fig. 8). This is the reason that in middle latitudes the weather tends to come from the west. It is also the reason that there is a westerly 'jet stream' at about 12 km, and that there is a significant difference in journey times flying eastward and westward across the North Atlantic and North Pacific, for example.

The second important feature of the temperature contrast is that it provides a source of potential energy. To see this, consider a tank of fluid with warm, light fluid separated from cold, dense fluid by a barrier as in Fig. 9a.

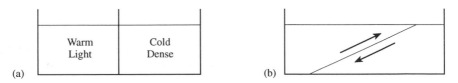

Fig. 9. (a) A tank with warm, light fluid and cold, dense fluid separated by a barrier. (b) What happens when the barrier is removed

If the barrier is removed, the fluids begin to move. Warm, light fluid starts to rise and move over the cold dense fluid which sinks and moves under it (Fig. 9b). The kinetic energy for the motion comes from the potential energy of the temperature contrast. In the Earth's atmosphere, the simple overturning does not occur because of the Coriolis force. Instead, the potential energy is converted into the kinetic energy of the weather systems with their geostrophic winds that we have already looked at. The battle between the warm air going towards the pole and the cold air going towards the Equator takes place in the warm and cold fronts in these weather systems.

As well as being important for our daily weather, these systems play an important role in climate. They transport heat towards the poles, stopping the high latitudes becoming too cold and the low latitudes too warm. They are also important in stopping the huge winds that we saw would occur if a ring of air moved from the Equator to our latitude.

Mathematics plays a crucial role in the theories of weather and climate. The predictions of tomorrow's weather and of climate change due to increased carbon dioxide are performed using mathematical models of the atmosphere. These models are solved on the fastest supercomputers that are available.

Solutions to Worksheet 1

1. If the top of your head is approximately a rectangle of sides 18 cm and 12 cm, its area is
$$A = 0.18 \times 0.12 = 0.0216\,\text{m}^2.$$
   Then the weight is
$$10^4 \times A = 216\,\text{kg}.$$

2. The Earth turns on its axis in
$$2\pi/\Omega = 86165\,\text{s}. \text{ This is called the sidereal day.}$$
   The solar day is $24 \times 3600 = 86400\,\text{s}$ which is 235 s, i.e. nearly 4 mins, longer.

3. I live at 52°N and I will use this in the solutions. You use your own latitude.
   At latitude 52°, $R\Omega \cos\phi = 286\,\text{ms}^{-1}$.

4. At latitude 52°, $R\Omega(\sin\phi)^2/\cos\phi = 469\,\text{ms}^{-1}$.
   This is 40% larger than the speed of sound!

5. The Coriolis force is from north to south.
   Its magnitude is $2\Omega(\sin\phi)V = 2.3 \times 10^{-3}\,\text{ms}^{-2}$.

Solutions to Worksheet 2

1. The pressure force is in the northward direction, as in Fig. 5a. The pressure difference is 15 mb = 1500 Pa in 400 km = 4×10^5 m. The magnitude of the pressure force is therefore
$$\frac{p_H - p_L}{\rho d} = \frac{1500}{1.2 \times 4 \times 10^5} = 3.125 \times 10^{-3}\,\text{ms}^{-2}.$$

2. $(p_H - p_L)/\rho d = 2\Omega(\sin \phi)V = 2.3 \times 10^{-3}$ ms^{-1} at latitude 52°. Therefore

$$p_H - p_L = 2.3 \times 10^{-3} \times (\rho \times d) = 1104 \text{ Pa}.$$

The pressure decrease is about 11 mb.

3. $(p_H - p_L)/\rho d = 2\Omega(\sin \phi)V.$

$$p_H - p_L = 22 \text{ mb} = 2200 \text{ Pa}, \qquad \rho = 1.2 \text{ kg m}^{-3}, \qquad d = 4 \times 10^5 \text{ m}.$$

$$V = \frac{2200}{1.2 \times 4 \times 10^5 \times 2\Omega \sin \phi}.$$

At latitude 52°, $V = 40$ ms^{-1}.

4.

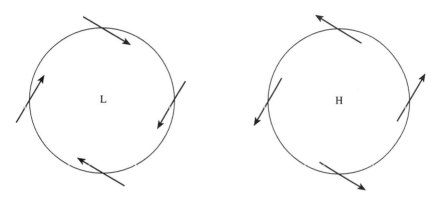

Further Reading

Unfortunately there is very little material at this level. Figure 1 is taken from *Met Matters*, a pack primarily for teachers, which deals with a thundery day and the Burns Day storm. It is available from Ross Reynolds, Dept. of Meterology, University of Reading, Reading RG6 2AU.

Now out of print, but perhaps available in libraries is:

R. P. Pearce, *The Observers book of weather* (F. Warne, 1980)

Some suitable material may be found in:

R. G. Barry and R. J. Chorley, *Atmosphere, weather and climate* (Routledge, 1992)

5 Square Roots and Seventeengons

by FRANCES KIRWAN

Balliol College, University of Oxford

1. Regular polygons

This masterclass concerns a very old problem in mathematics which goes back to the ancient Greeks and was not solved until the nineteenth century; indeed, as we shall see, even today some questions remain unanswered. The solution involves some beautiful algebra which is often covered in second or third year university courses, but the problem itself is geometric and easy to explain.

Since the earliest days of mathematics people have constructed geometric shapes using the most basic tools of geometry: a straight-edge for drawing straight lines and a pair of compasses for drawing circles (or, in their more primitive forms, any straight object and a drawing implement attached to a piece of string). The ancient Greeks (who were very sophisticated mathematicians) were particularly skilled at this; they were able to construct all sorts of different shapes using only these implements. They also invented much more complicated mechanisms to enable them to perform a wider range of constructions, but they preferred if possible to use a straight-edge and compasses alone.

The rules of the game are that you start with a fixed line segment of unit length, and you are allowed to draw *lines* through any pairs of points you choose with your straight edge (or ruler), and *circles* with centres wherever you like with your compasses, having measured the radius if you wish against any line segment already drawn.

Example 1
To draw a regular hexagon, first draw a circle and then, without altering the spread of your compasses, place the point of the compasses on the circumference of the circle and mark two points on the circumference. Next place the point of the compasses at each of these two points in turn and repeat the process to obtain two new points on the circumference. Repeat one more time and you will have six points spaced around the circumference of the circle. Join these by straight lines to obtain a regular hexagon (see Fig. 1).

Example 2
One way to draw an equilateral triangle (that is, a triangle with three equal sides) is to construct six points equally spaced around a circle, as in Example 1, and then join up every second point.

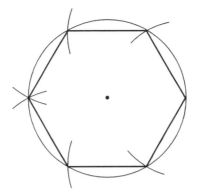

Fig. 1. Construction of a regular hexagon

Example 3

One way to draw a square is to first draw a circle and then draw a diameter of the circle (that is, a line segment through the centre which ends at two opposite points on the circumference of the circle). Let us call the endpoints of the diameter P and Q. Now widen your compasses a little and draw two more circles with centres at P and Q (these circles should be the same size as each other but a bit bigger than the first circle). The two bigger circles will meet each other at two points; by the symmetry of the situation the line through these points will meet the diameter PQ at right-angles at its midpoint (that is, at the centre of the first circle) and will meet the first circle at two points which we can call R and S. Then the four points P, R, Q and S will be evenly spaced around the circumference of the first circle and can be joined with straight lines to form a square (see Fig. 2).

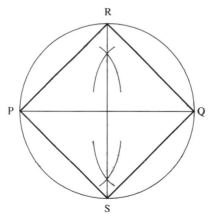

Fig. 2. Construction of a square

Definition 1

A polygon (that is, a straight-sided figure) with n sides of equal length is called a *regular polygon* or, if it is desirable to specify the number of sides, a *regular n-gon*. For some values of n we have more familiar names for n-gons.

> A regular 3-gon is an equilateral triangle.
> A regular 4-gon is a square.
> A 5-gon is also called a pentagon.
> A 6-gon is also called a hexagon.
> A 7-gon is also called a heptagon.
> An 8-gon is also called an octagon.

We have seen how to construct regular n-gons with a straight-edge and compasses for $n = 3, 4, 6$. What about $n = 5$? The construction is possible but much more complicated.

2. Construction of a regular pentagon

To construct a regular pentagon with its corners on a given circle centre O we can proceed as follows (the construction is illustrated in Fig. 3).

(i) Choose any point P_1 on the circumference of the circle.
(ii) Using the method of Example 3 (see also Exercise 3 on Worksheet 1) draw a line segment through O at right-angles to OP_1 meeting the circle at B.
(iii) Using the method of Example 3 again, construct the midpoint D of OB.
(iv) Using the method given in Exercise 2 of Worksheet 1, construct the line which bisects the angle ODP_1 (that is, it divides the angle into two equal parts). Let N be the point where this line meets the line segment OP_1.
(v) Using the method of Example 3 yet again, draw a line segment through N at right-angles to ON meeting the circle at a point P_2.

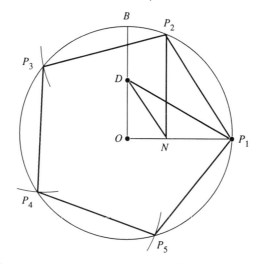

Fig. 3. Construction of a regular pentagon

(vi) Spread your compasses to the distance between P_1 and P_2 and use them to mark three more points P_3, P_4 and P_5 so that P_1, P_2, P_3, P_4 and P_5 are the same distance apart around the circumference of the circle. These five points are then the corners of a regular pentagon (see Fig. 3).

How can we show that this construction works? What we must do is show that the angle P_1OP_2 of the triangle OP_1P_2 at O is

$$360/5 = 72$$

degrees. For this we need one of the most famous theorems proved by the ancient Greeks.

Theorem 1 (Pythagoras' theorem)
If a, b, c are the lengths of the sides of a right-angled triangle and c is the length of the hypotenuse (that is, the longest side) then

$$c^2 = a^2 + b^2.$$

Proof
Consider the two ways of dividing up a square of side $a + b$ shown in Fig. 4. Each contains four identical triangles, but the first contains in addition one square of side c, whereas the second contains in addition one square of side a and one square of side b. Therefore the area of the square of side c must be equal to the sum of the areas of the squares of sides a and b. □

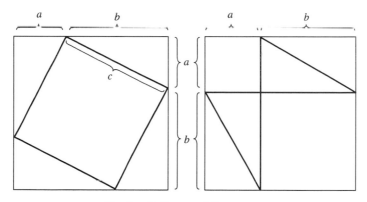

Fig. 4. Pythagoras' theorem

We also need to know about similar triangles. Recall that two triangles ABC and DEF are called *similar* if they have the same shape but perhaps different sizes (see Fig. 5). Then the corresponding angles of the two triangles are the same (for example the angle BAC is equal to the angle EDF) and the ratios of the lengths of the corresponding sides are the same; that is

$$\frac{AB}{DE} = \frac{BC}{EF} = \frac{AC}{DF}.$$

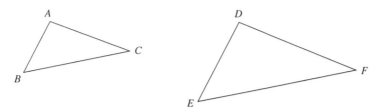

Fig. 5. Similar triangles

Proof that the construction of the regular pentagon works

To make our calculations a little easier we can assume that the circle in which the pentagon is to be drawn has radius 2 units. Consider Fig. 6; we wish to calculate the length y of ON. The length of OP_1 is 2 units and the length of OD is 1 unit (it is half the radius). Therefore by Pythagoras' theorem (Theorem 1 above) the length of DN is $\sqrt{5}$. Since the line DN bisects the angle ODP_1, if we draw NM meeting DP_1 at right-angles at M then the triangles DNO and DNM are reflections of each other in the line DN. Thus the length of NM is y and the length of DM is 1 so the length of MP_1 is $\sqrt{5} - 1$. But the two right-angled triangles ODP_1 and MNP_1 are similar so

$$\frac{y}{1} = \frac{\sqrt{5} - 1}{2}.$$

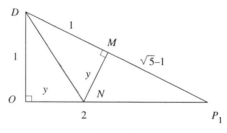

Fig. 6

Now consider the first triangle in Fig. 7. In this diagram the large triangle has the same angles as the lower of the two small triangles. Therefore they are similar triangles and so we find, comparing the lengths of their sides, that

$$\frac{x}{2} = \frac{2 - x}{x}.$$

Multiplying through by $2x$ we get

$$x^2 = 2(2 - x) = 4 - 2x$$

and adding $2x$ to each side gives

$$x^2 + 2x = 4.$$

Then
$$(x+1)^2 = x(x+1) + 1(x+1) = x^2 + 2x + 1 = 5;$$
so taking square roots we get
$$x + 1 = \pm\sqrt{5}.$$
Since x is positive we must have the positive square root here, so $x = \sqrt{5} - 1$. But we found above that $y = (\sqrt{5} - 1)/2$, so
$$2y = \sqrt{5} - 1 = x.$$
Thus comparing the two diagrams in Fig. 7 we find that the angle NOP_2 is 72 degrees, which is what we wanted. $\square$

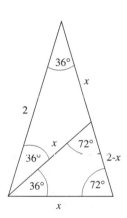
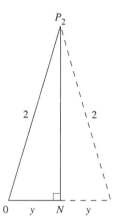

Fig. 7

Remark 1
The square root which has appeared in this proof will be significant later.

The ancient Greeks knew all these constructions. They called the ratio $1/y$, where as above $y = (\sqrt{5} - 1)/2$, the *golden ratio*, because they thought that rectangles whose sides were in this ratio were the most aesthetically pleasing. The reason was that if a square is cut off a rectangle whose sides are in this ratio then the sides of the remaining rectangle are in the same ratio (see Fig. 8).

The Greeks tried very hard to find a construction of a regular heptagon (7-gon) using only a straight-edge and compasses, and so did later mathematicians, but they failed. They also tried very hard to find a method for trisecting (that is, dividing into three equal parts) any given angle, which would have enabled them to construct a regular nonagon (9-gon). In fact neither of these constructions is possible, but nobody seems to have even guessed this until the eighteenth century and it was not proved until the nineteenth century.

Thus we have two interesting questions about regular polygons to consider.

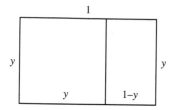

Fig. 8. The golden ratio

Question 1
For which values of n can a regular n-gon be constructed using straight-edge and compasses alone?

Question 2
If it is not possible to construct a regular n-gon using only a straight-edge and compasses, how do you *prove* this (to stop people wasting their time trying to do it)?

Worksheet 1

1. Prove that the construction of a regular hexagon in Example 1 works, by showing that the centre of the circle, together with the two points first constructed on the circumference, form the vertices of a triangle which is equilateral and therefore has angles all equal to $60 = 360/6$ degrees.

   Now do the construction, and also construct an equilateral triangle and a square using the methods described in the text.

2. Show that any given angle POQ can be bisected by the following procedure which uses only a straight-edge and compasses.

   First draw a circle centre O meeting the line segments OP and OQ at points A and B. Then draw two circles of the same radius with centres at A and B. If these circles meet at C show that the line segment OC bisects the angle POQ.

3. Show that it is possible to construct the perpendicular bisector of any line segment PQ (that is, the line which meets PQ at right-angles at the midpoint of PQ) in the following way.

   Draw two circles of the same radius with centres at P and Q, making sure that the radius is large enough so that the circles meet in two points A and B. Show that the line joining A and B is the perpendicular bisector of PQ.

   How can you use this idea to construct the line which meets PQ at right-angles at P?

4. Use the last exercise to show that it is possible to drop a perpendicular onto a line L from any point R not lying on L; that is, to construct the line through

R which meets *L* at right-angles. (You may find it helpful to draw a circle with centre at *R* which is large enough to meet the line *L* at two points, *P* and *Q*).

5. Show that if it is possible to construct line segments of lengths *x* and *y* using only a straight-edge and compasses, then it is also possible to construct the point in the plane with coordinates (x, y). Here we take coordinates in the usual way so that the given line segment of unit length is the line segment on the *x*-axis joining the origin $(0,0)$ to the point $(1,0)$.

Conversely show that if it is possible to construct the point (x, y) then it is possible to construct line segments of lengths *x* and *y*.

6. Show that if $a \neq 0$ then

$$\left(x + \frac{b}{2a}\right)^2 = x\left(x + \frac{b}{2a}\right) + \frac{b}{2a}\left(x + \frac{b}{2a}\right) = x^2 + \frac{bx}{a} + \frac{b^2}{4a^2}.$$

Deduce that if *x* satisfies the quadratic equation

$$ax^2 + bx + c = 0$$

where $a \neq 0$ then

$$a\left(x + \frac{b}{2a}\right)^2 = \frac{b^2 - 4ac}{4a}.$$

Divide both sides by *a* and take square roots to conclude that

$$x = \frac{-b \pm \sqrt{b^2 - 4ac}}{2a}.$$

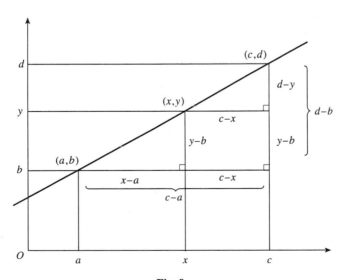

Fig. 9

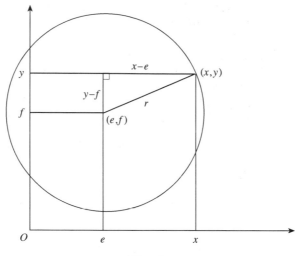

Fig. 10

7. With the help of Fig. 9, show that if a point with coordinates (x, y) lies on the line L through two points (a, b) and (c, d), then
$$(d - b)(x - a) = (c - a)(y - b).$$
With the help of Fig. 10, use Pythagoras' theorem (Theorem 1 above) to show that if (x, y) lies on the circle C with centre (e, f) and radius r, then
$$(x - e)^2 + (y - f)^2 = r^2.$$

8. Show that
$$(c - a)(y - f) = (c - a)(y - b) + (c - a)(b - f)$$
for any y, a, b, c and f. Now suppose that the point (x, y) lies on both the line L and the circle C described in the last Exercise. Use that Exercise to show that
$$(c - a)^2(x - e)^2 + ((x - a)(d - b) + (c - a)(b - f))^2 = (c - a)^2 r^2.$$
Using Exercise 6, deduce that x (and hence also y) can be expressed in terms of a, b, c, d, e, f and r using only addition, subtraction, multiplication, division and square roots.

[N.B. There is no need to work out the actual formulas for x and y in terms of a, b, c, d, e, f, and r.]

9. Show that if (x, y) lies on the circle C of Exercise 7, and also on another circle D of centre (h, k) and radius s, then
$$x^2 - 2ex + e^2 + y^2 - 2fy + f^2 = r^2$$
and
$$x^2 - 2hx + h^2 + y^2 - 2ky + k^2 = s^2$$
so
$$2(h - e)x + 2(k - f)y = r^2 - s^2 + h^2 - e^2 + k^2 - f^2.$$

As in Exercise 8 deduce that x and y can be expressed in terms of e, f, r, h, k and s using only addition, subtraction, multiplication, division and square roots (but do not work out the actual formulas).

3. Constructible numbers and square roots

In order to answer the questions at the end of the last section, we need to investigate which points in the plane can be constructed using only a straight-edge and compasses, starting from a given line segment of unit length. Recall from Exercise 5 of Worksheet 1 that a point with coordinates (x, y) can be constructed if and only if it is possible to construct line segments of lengths x and y. This means that the following definition will be useful.

Definition 2
Let us draw a line through our given line segment of length one unit, marking one end of the line segment as 0 and the other end as 1. (If we wish, we can think of this line as the x-axis of a coordinate system as in Exercise 5 of Worksheet 1). Then a number a will be called *constructible* if the point on this line whose distance from 0 is a can be constructed using only a straight-edge and compasses.

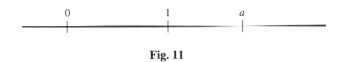

Fig. 11

Remark 2
As usual, positive numbers are represented by points on the line which are the same side of 0 as 1 (see Fig. 11), and negative numbers are represented by points on the opposite side of 0 from 1.

Now we have a very important fact about constructible numbers.

Theorem 2
If a and b are constructible then so are $a + b$, $a - b$, ab, a/b (provided that $b \neq 0$) and $\sqrt{a}$ (provided that a is positive).

Proof
We can assume that the points a and b have already been marked on the line through 0 and 1. To construct $a + b$ and $a - b$, we spread our compasses to the length of the line segment joining 0 and b and draw a circle of this radius with its centre at a. This circle meets the line in the two points $a + b$ and $a - b$ (see Fig. 12 for the case when b is positive).

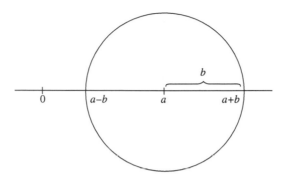

Fig. 12.   Construction of $a \pm b$

In order to construct ab and a/b (if $b \neq 0$), let us assume that a and b are both positive; it is enough just to deal with this case because we know that if a and b are constructible then so are $-a$ and $-b$, and also that if $-ab$ and $-a/b$ are constructible then so are ab and a/b. Now we draw another line through 0, use the compasses to mark the points with distance 1, a and b from 0 on this line, and label these points I, A and B (see Fig. 13).

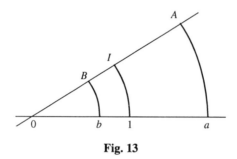

Fig. 13

To construct the point ab, first we draw the line segment $1B$ joining 1 to B, and then the line through 1 perpendicular to this line segment (see Exercise 3 of Worksheet 1). Next we drop a perpendicular from a onto this line through 1 (see Exercise 4 of Worksheet 1) and extend it until it meets the line $0B$ extended, at a point of P, say (see Fig. 14). The line segments aP and $1B$ are parallel (because there is a line perpendicular to both of them), so the triangles $aP0$ and $1B0$ are similar. Thus the ratios of the lengths $0P$ to $0B$ and of the lengths $0a$ to 01 are equal: that is,

$$\frac{0P}{b} = \frac{a}{1} = a.$$

so the length of $0P$ is ab.

To construct a/b, we first join b to I, next drop a perpendicular from a to the line joining b and I, and then construct a line through a at right-angles to this perpendicular bisector, meeting the line $0I$ at Q, say (see Fig. 15). The

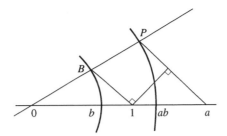

Fig. 14. Construction of ab

triangles $0aQ$ and $0bI$ are similar, so the ratios of the lengths $0Q$ to $0I$ and of the lengths $0a$ to $0b$ are equal; that is,

$$\frac{0Q}{1} = \frac{a}{b},$$

so the length of $0Q$ is a/b.

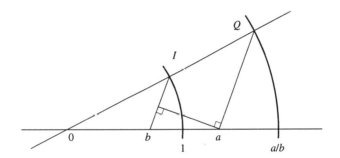

Fig. 15. Construction of a/b

Finally in order to construct $\sqrt{a}$ if a is positive, we first construct $a + 1$ as above and then the midpoint $(a + 1)/2$ of the line segment joining 0 and $a + 1$. Now we can construct the circle with centre at $(a + 1)/2$ and radius $r = (a + 1)/2$, which passes through 0 and $a + 1$. Next we construct the line through

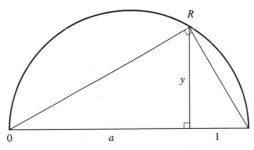

Fig. 16. Construction of $\sqrt{a}$

a perpendicular to the line through 0 and a (see Exercise 3 of Worksheet 1 again), and mark the point, R say, where it meets the circle (see Fig. 16).

The line segments $0R$ and $R(a + 1)$ meet at right-angles at R, because in a circle the angle subtended by any diameter is a right angle (look at Fig. 17 and remember that the angles of a triangle always add up to 180 degrees).

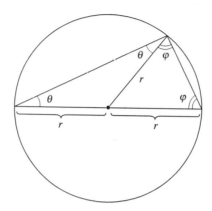

Fig. 17.   The angle subtended by a diameter is a right angle

Since the two smaller angles of any right-angled triangle add up to 90 degrees, it follows that the three triangles in Fig. 16 are all similar, and so the length y of the line segment Ra satisfies

$$\frac{y}{1} = \frac{a}{y}.$$

Therefore $y^2 = a$ and so $y = \sqrt{a}$.   □

From this theorem together with Exercises 8 and 9 of Worksheet 1, we obtain the following characterization of constructible numbers.

Theorem 3
The constructible numbers are precisely those which can be obtained from the integers by repeated applications of square roots, addition, subtraction, multiplication and division.

Proof
Any integer can be obtained from 0 and 1 by repeated applications of addition or subtraction. Therefore Theorem 2 tells us that any number which can be obtained from the integers by repeated applications of square roots, addition, subtraction, multiplication and division is constructible. Conversely Exercises 8 and 9 of Worksheet 1 tell us that any constructible number can be obtained from the integers by repeated applications of square roots, addition, subtraction, multiplication and division, because at each stage of the construction the new

points constructed are the intersections of lines through points which have already been constructed and/or circles whose centres and radii have already been constructed.   □

Remark 3
Recall that in our construction of a regular pentagon the crucial length was $(\sqrt{5} - 1)/2$. This of course fits with Theorem 3.

This theorem gives us an answer, and quite a useful answer, to the question of which numbers are constructible. In particular, we now know that every rational number is constructible. In case you have not already come across the concept of a rational number, here is the definition.

Definition 3
A number is *rational* if it can be expressed in the form n/m where n and m are integers and $m \neq 0$.

The answer given by Theorem 3 to the question of which numbers are constructible is not entirely satisfactory, however. For example, if we are given a number, such as the cube root $\sqrt[3]{2}$ of 2, which does not *seem* to be expressible purely in terms of the integers and repeated applications of square roots, addition, subtraction, multiplication and division, how can we be sure that it is not given by some extremely complicated expression in these terms? Could perhaps *all* numbers be constructible? Or at the opposite extreme, could perhaps only rational numbers be constructible?

Question 3
Are all constructible numbers rational?

Question 4
Are all numbers constructible?

The answer in each case is, in fact, negative. For example $\sqrt{2}$ is constructible (by Theorem 3) but not rational (the proof is given below), and $\sqrt[3]{2}$ is not constructible (we shall discuss this further, though without a complete proof, in Section 4).

A proof that $\sqrt{2}$ is not a rational number was first discovered by the ancient Greeks, and caused great consternation! Until then, the Greeks had assumed that all 'measurable quantities' (including $\sqrt{2}$ as the length of the diagonal of a unit square) were rational.

Theorem 4
$\sqrt{2}$ is not a rational number.

Proof
We shall show that it is impossible for $\sqrt{2}$ to be a rational number by showing

that the assumption that it *is* rational leads to a contradiction. So suppose, in order to get a contradiction, that

$$\sqrt{2} = n/m$$

where n and m are positive integers. If n and m are both even, then we can divide them both by 2 without changing the ratio n/m. We can repeat this until at least one of n and m is odd. So we can now assume that

$$\sqrt{2} = n/m$$

where n and m are positive integers, *not both even*. Then

$$2 = n^2/m^2$$

so

$$n^2 = 2m^2$$

so n^2 is even. Therefore n must be even (if n were odd then n^2 would be odd too, because an odd number times an odd number is odd). This means that we can write

$$n = 2p$$

for some integer p. Then

$$2m^2 = n^2 = 4p^2$$

so dividing through by 2 we get

$$m^2 = 2p^2$$

and hence m^2 is even, so m is even (by the argument used above for n). Thus we have shown that n and m are both even and this contradicts our assumption. This contradiction shows that it is impossible for $\sqrt{2}$ to be a rational number. □

Theorem 3 has turned our geometric problem of finding which points in the plane can be constructed from a given line segment of unit length using only a straight-edge and compasses into an algebraic problem. In the final two sections we shall discuss this algebraic problem and apply it to our original questions concerning the construction of regular polygons (Questions 1 and 2 at the end of Section 2).

Worksheet 2

1. Show that if m divides n (this means that n is equal to the product of m and some integer k) and it is possible to construct a regular n-gon using straight-edge and compasses, then it is possible to construct a regular m-gon by joining up some of the vertices of the regular n-gon.
2. Show that if it is possible to construct a regular n-gon using straight-edge and compasses, then it is possible to construct
   (i) a circle passing through all its vertices, and
   (ii) an angle of $360/n$ degrees.

Conversely show that if it is possible to construct an angle of $360/n$ degrees, then it is possible to construct a regular n-gon.

3. Use Exercise 2 of Worksheet 1 and Exercise 2 above to show that if a regular n-gon is constructible then so is a regular $2n$-gon. Illustrate this by constructing a regular 12-gon.

4. Suppose that n and m are positive integers with no common factors (that is, there is no positive integer other than 1 which divides both n and m). Then we can use the *Euclidean algorithm* to find integers a and b satisfying.

$$an + bm = 1.$$

It works like this. We may as well assume that $n > m$. Then we can divide n by m to give q, say, with remainder r; that is, we have

$$n = qm + r$$

where $0 \leq r < m$. If $r = 0$ then m divides n, and because n and m have no common factors this means $m = 1$. Of course we can then take $a = 0$ and $b = 1$ to get $an + bm = 1$. Otherwise $r \neq 0$. Using the equation $n = qm + r$, show that any common factor of m and r is a common factor of n and m, and deduce that m and r can have no common factors. Now we can repeat the procedure, replacing n by m and replacing m by r, to find integers q_1 and r_1 satisfying

$$m = q_1 r + r_1$$

and $0 \leq r_1 < r$. If $r_1 = 0$ then r divides m, and because m and r have no common factors this means $r = 1$. Check that in this case we can take $a = 1$ and $b = -q$ to get $an + bm = 1$. Otherwise $r_1 \neq 0$. Show that

$$r_1 = m - q_1 r = m - q_1(n - qm) = (1 + q_1 q)m - q_1 n.$$

Using the equation $m = q_1 r + r_1$, show that any common factor of r and r_1 is a common factor of m and r, and deduce that r and r_1 can have no common factors. Now we can repeat the procedure again, replacing m by r and replacing r by r_1, to find integers q_2 and r_2 satisfying

$$r = q_2 r_1 + r_2$$

and $0 \leq r_2 < r_1$. Using the argument above, show that if $r_2 = 0$ then $r_1 = 1$ and deduce that we can take $a = -q_1$ and $b = 1 + q_1 q$ to get $an + bm = 1$.

We can continue to repeat this procedure until we reach a remainder which is 0 (this must happen in at most m steps since each remainder is a non-negative integer strictly less than the remainder before). Each remainder in the procedure can be expressed in the form $an + bm$ for some integers a and b, by backwards substitution. Once we reach a remainder which is 0, the previous remainder must be 1, and therefore 1 can be expressed in the form $an + bm$ as required.

Now apply this procedure to find integers a and b satisfying

$$a \times 25 + b \times 57 = 1.$$

5. Suppose that we can construct angles of of $360/n$ degrees and of $360/m$ degrees using only a straight-edge and compasses. Show that we can construct an angle of

$$a(360/m) + b(360/n)$$

degrees, for any integers a and b such that $a(360/m) + b(360/n)$ is positive.

6. Suppose that n and m are positive integers with no common factors, so that by Exercise 4 we can find integers a and b such that $an + bm = 1$. Suppose also that it is possible to construct a regular n-gon and a regular m-gon using only a straight-edge and compasses. Use Exercises 2 and 5 to show that it is then possible to construct an angle of $360/nm$ degrees, and hence also a regular nm-gon.

Illustrate this when $n = 3$ and $m = 5$ by constructing a regular 15-gon on Fig. 18.

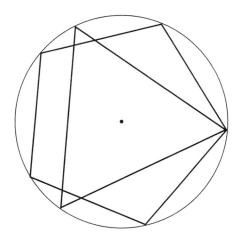

Fig. 18.   Construction of a regular 15-gon

7. A prime number is an integer $p > 1$ which has no (positive integer) factors other than itself and 1. A *Fermat prime* (called after the seventeenth century French mathematician Fermat) is a prime number of the form

$$p = 2^{2^r} + 1$$

for some non-negative integer $r = 0, 1, 2, 3, \ldots$. What are the first three Fermat primes? Roughly how many Fermat primes less than one million do you think there could be?

8. In fact (see Section 5) it is possible to construct a regular n-gon using only a straight-edge and compasses *if and only if* n is of the form

$$n = 2^a pq \cdots r$$

where a is either zero or a positive integer and $p, q, \ldots, r$ are distinct Fermat primes (there may be any number of them from none upwards but no repeats are allowed). Which values of n less than or equal to 50 are of this form? Do you expect there to be more or fewer numbers of this form between 50 and 100 than there are less than 50?

4. Polynomial equations

How can we use Theorem 3 to decide whether a given number α is constructible? Of course this is straightforward if α can easily be expressed in terms of the integers and taking square roots, addition, subtraction, multiplication and division, but what if α is not obviously given by such an expression? First of all, it turns out that if α is constructible it must satisfy a polynomial equation with rational coefficients which is not identically zero. (Not all numbers do satisfy such equations: for example the number π does not, nor does the base e for the natural logarithms).

Example 4
The constructible number

$$\beta = \sqrt{2 + \sqrt{17}} - 1$$

satisfies

$$\left((\beta + 1)^2 - 2 \right)^2 = 17$$

or equivalently

$$\beta^4 + 4\beta^3 + 2\beta^2 - 4\beta - 16 = 0.$$

Note that this polynomial equation has degree $4 = 2 \times 2$.

Example 5
The constructible number

$$\gamma = \sqrt{\sqrt{2} + \sqrt{3}}$$

satisfies

$$\gamma^2 = \sqrt{2} + \sqrt{3},$$
$$\gamma^4 = 5 + 2\sqrt{6},$$

and

$$\gamma^8 = 49 + 20\sqrt{6},$$

so

$$\gamma^8 - 10\gamma^4 + 1 = 0.$$

Note that this polynomial equation has degree $8 = 2 \times 2 \times 2$.

If a number α does satisfy a polynomial equation with rational coefficients which is not identically zero then it will satisfy many such equations, but we can choose one with the smallest possible degree. It turns out that if this degree is a power of two then α is constructible (recall that a power of two is two multiplied by itself a certain number of times; note that by convention two multiplied by itself zero times is one, so one is a power of two) and otherwise α is not constructible. Unfortunately the proof of this fact is too complicated to explain here; it is given in Stewart (1989) but be warned – this is a book written primarily for university students! The proof uses a part of mathematics called Galois theory, after the French mathematician Évariste Galois, who was killed in a duel in 1832 a few months before his twenty-first birthday. The night before the duel Galois wrote to a friend outlining his ideas on the solutions of polynomial equations, but it was not until more than a decade later that the mathematical community realized the importance of what he had done. Chapter 8 of Stewart (1987) contains an informal introduction to the main ideas of Galois theory.

Example 6

The cube root $\sqrt[3]{2}$ of 2 is not constructible because it satisfies the equation $x^3 - 2 = 0$ of degree 3, but does not satisfy any polynomial equation with rational coefficients of degree 1 or 2. This demonstrates the impossibility of one of the famous constructions which the ancient Greeks tried and failed to perform using straight-edge and compasses alone: duplication of the cube, or in other words the construction of a cube with twice the volume of a given cube.

How can we apply this to the question of which regular polygons are constructible using only a straight-edge and compasses? A regular n-gon can be constructed if and only if an angle of $360/n$ degrees can be constructed (see Exercise 2 of Worksheet 2), or equivalently when it is possible to construct the sides of a right-angled triangle with hypotenuse (longest side) of length 1 and one angle of $360/n$ degrees. Therefore a regular n-gon is constructible if and only if the lengths of the sides of such a triangle satisfy polynomial equations with rational coefficients which are not identically zero, *and* the smallest possible degrees of such equations are powers of two.

Example 7
Consider a right-angled triangle with hypotenuse of length 1 and one angle equal to $360/7$ degrees. Let α be the length of the side adjoining the angle of $360/7$ degrees which is not the hypotenuse. Then (see below) α satisfies the polynomial equation
$$8\alpha^3 + 4\alpha^2 - 4\alpha - 1 = 0$$
of degree 3 (which is not a power of two), and α satisfies no polynomial equations with rational coefficients and degree 1 or 2. This is why it is impossible to construct a regular heptagon (7-gon) using only a straight-edge and compasses.

In fact anyone who has studied sines and cosines will know that $\alpha = \cos(360/7)^\circ$. The equation above satisfied by α can be deduced from the fact that if $\theta = (360/7)°$ then $\cos 3\theta = \cos 4\theta$ and $\alpha = \cos\theta \neq 1$. If α also satisfied a polynomial equation with rational coefficients and degree 1 or 2, then the polynomial $8x^3 + 4x^2 - 4x - 1$ would have to factorize as a product of two polynomials with rational coefficients and degrees 1 and 2, and thus would have to have a rational root, p/q say, where p and q are integers with no common factors. Substituting $x = p/q$ and multiplying through by q^2 we would get

$$8p^3 + 4p^2q - 4pq^2 - q^3 = 0$$

which would imply that q^3 was divisible by p and $8p^3$ was divisible by q. Since p and q have no common factors, the only way this could happen would be if $p = \pm 1$ and $q = \pm 1, \pm 2, \pm 4$ or ± 8, but it is easy to check that none of these possibilities gives a solution to the equation.

Galois theory is concerned with the solutions of polynomial equations, and hence involves the use of complex numbers. A *complex number* is an expression of the form

$$a + b\sqrt{-1}$$

where a and b are real numbers (that is, the familiar sort of numbers we use for measurements, etc.) and $\sqrt{-1}$ is an 'imaginary number' which is a square root of -1 (see e.g. Chapter 11 of Stewart (1987), Chapter II(§5) of Courant and Robbins (1941) or Chapter 6 of Coxeter (1969)). We call a the real part of $a + b\sqrt{-1}$ and b the imaginary part. Complex numbers are added and multiplied together by the rules

$$(a + b\sqrt{-1}) + (c + d\sqrt{-1}) = (a + c) + (b + d)\sqrt{-1}$$

and

$$(a + b\sqrt{-1})(c + d\sqrt{-1}) = (ac - bd) + (bc + ad)\sqrt{-1}.$$

The reason that complex numbers are important in the study of polynomial equations is that *every* polynomial equation has solutions if complex numbers are allowed (even if the coefficients of the polynomial are themselves complex numbers). This fact is sometimes called the fundamental theorem of algebra.

We can represent a complex number $a + b\sqrt{-1}$ by the point in the plane with coordinates (a, b). Let us call $a + b\sqrt{-1}$ constructible if the corresponding point (a, b) in the plane is constructible from the line segment joining the origin to $(1, 0)$ using only a straight-edge and compasses. Recall from Exercise 5 of Worksheet 1 that (a, b) is constructible if and only if a and b are both constructible numbers. Since $\sqrt{-1}$ is a square root of an integer, Theorem 3 is still valid if complex numbers are allowed. So is the condition for constructible numbers given earlier in this section: a complex number is constructible if and only if, firstly, it satisfies a polynomial equation with rational coefficients which is not identically zero and, secondly, the smallest possible degree of such an equation is a power of two.

The modulus $|a + b\sqrt{-1}|$ of a complex number $a + b\sqrt{-1}$ is the distance

$\sqrt{a^2 + b^2}$ of (a, b) from the origin, and if $a + b\sqrt{-1}$ is not equal to 0 then its argument $\arg(a + b\sqrt{-1})$ is the angle between the line segments joining $(1, 0)$ and (a, b) to the origin, measured in an anticlockwise direction (see Fig. 19). It follows from the formula for the multiplication of complex numbers and some elementary trigonometry that

$$|(a + b\sqrt{-1})(c + d\sqrt{-1})| = |a + b\sqrt{-1}||c + d\sqrt{-1}|$$

and

$$\arg((a + b\sqrt{-1})(c + d\sqrt{-1})) = \arg(a + b\sqrt{-1}) + \arg(c + d\sqrt{-1})$$

provided we interpret the last equality as an equality between angles so that, for example, $270° + 180° = 90°$ is allowed.

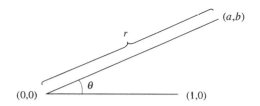

Fig. 19. $r = |a + b\sqrt{-1}|$ and $\theta = \arg(a + b\sqrt{-1})$

For any positive integer $n > 4$ let ω_n denote the complex number whose modulus is one and whose argument is $360/n$ degrees (so that its real and imaginary parts are the lengths of the sides of a right-angled triangle with hypotenuse of length 1 and one angle of $360/n$ degrees). Then it is possible to construct a regular n-gon with only a straight-edge and compasses if and only if ω_n is a constructible complex number. Using the equations above we find easily that ω_n is a root of the polynomial equation $x^n - 1 = 0$. Since $\omega_n \neq 1$ and

$$x^n - 1 = (x - 1)(x^{n-1} + x^{n-2} + \cdots + x + 1)$$

it follows that ω_n is a root of the polynomial $x^{n-1} + x^{n-2} + \cdots + x + 1$.

First consider the case when n is a prime number p. It can be shown that ω_p satisfies no nontrivial polynomial equations with rational coefficients and degree smaller than the degree of $x^{p-1} + x^{p-2} + \cdots + x + 1$, which is $p - 1$. We conclude that a regular p-gon can be constructed using only a straight-edge and compasses if and only if $p - 1$ is a power of two.

Now suppose that $n = p^2$ is the square of an odd prime number p. It can then be shown that ω_n is a root of the polynomial $x^{p(p-1)} + x^{p(p-2)} + \cdots + x^p + 1$ but is not a root of any polynomials with rational coefficients and smaller degree than $p(p - 1)$ which are not identically zero. We conclude that a regular p^2-gon can never be constructed, since $p(p - 1)$ contains an odd factor p so it cannot be a power of two. It then follows from Exercise 1 of Worksheet 2 that a regular n-gon can never be constructed if n is divisible by p^2 for some odd prime number p.

Remark 4

If $w > 1$ is an integer such that $p = 2^w + 1$ is prime, then w must itself be a power of two. For otherwise we would be able to write $w = uv$ where $v > 1$ is an *odd* integer and $u \geq 1$ is an integer, and then $p = 2^w + 1$ would factorize as

$$2^{uv} + 1 = (2^u + 1)(2^{u(v-1)} - 2^{u(v-2)} + \cdots - 2^u + 1),$$

contradicting the assumption that p is prime.

5. Fermat primes

At the end of the last section we concluded that if p is an odd prime then a regular n-gon can never be constructed when n is divisible by p^2, and a regular p-gon can be constructed if and only if p is a Fermat prime; that is, when

$$p = 2^{2^r} + 1$$

for some integer $r \geq 0$. Using Exercises 1, 3 and 6 of Worksheet 2 we obtain the following result which was stated in Exercise 8 of Worksheet 2.

Theorem 5 (Gauss)

It is possible to construct a regular n-gon using only a straight-edge and compasses *if and only if* n is of the form

$$n = 2^a pq \cdots r$$

where a is either zero or a positive integer and $p, q, \ldots, r$ are distinct Fermat primes. (There may be any number of Fermat primes here, including none, but they must be all different).

Proof

Let us write p^e to mean the product $p \times p \times \cdots \times p$ of e factors of p when p and e are positive integers. Then any positive integer $n > 1$ can be expressed in exactly one way as a product of prime numbers

$$n = p_1^{e_1} p_2^{e_2} \cdots p_k^{e_k}$$

where k and $e_1, e_2, \ldots, e_k$ are positive integers and $p_1, p_2, \ldots, p_k$ are prime numbers satisfying $p_1 < p_2 < \cdots < p_k$ (see e.g. Chapter 2 of Baylis and Haggarty (1988)). If a regular n-gon is constructible then, as we have seen in the last section, n is not divisible by the square of any odd prime so n must be of the form

$$n = 2^a pq \cdots r$$

where a is either zero or a positive integer and $p, q, \ldots, r$ are *distinct* primes. Moreover by Exercise 1 of Worksheet 2 the constructibility of a regular n-gon, when $n = 2^a pq \cdots r$, implies that it is possible to construct a regular p-gon, a regular q-gon and so on, which implies that $p, q, \ldots, r$ must be Fermat primes.

Conversely if n is of this form where $p, q, \ldots, r$ are distinct Fermat primes then it is possible to construct a regular p-gon, a regular q-gon and so on, which means by Exercise 6 of Worksheet 2 that it is possible to construct a regular $pq \cdots r$-gon, and therefore by Exercise 3 of Worksheet 2 it is possible to construct a regular n-gon.   □

The first five Fermat primes given by taking $r = 0, 1, 2, 3, 4$ are

$$p = 3, 5, 17, 257 \quad \text{and} \quad 65\,537$$

In particular a regular 17-gon can be constructed using a straight-edge and compasses (see below). This was first proved by the great German mathematician Gauss (1777–1855) while he was still a teenager, and the discovery is said to have pleased him so much that it convinced him to devote his life to mathematics.

It is still not known whether there are more than five Fermat primes. The number $2^{2^r} + 1$ is not prime for all values of r greater than or equal to five for which it has been tested.

6. Construction of the regular 17-gon using straight-edge and compasses

Here is a construction of a regular 17-gon using only a straight-edge and compasses (see Fig. 20). The construction is adapted from Chapter 17 of Stewart (1989) where it is possible to find a proof that this method does indeed give a regular 17-gon.

1. Draw a circle centre O, radius 1. Choose any point P_1 on its circumference to be the first vertex of the 17-gon.
2. Draw a line through O at right angles to OP_1 meeting the circle at B.
3. Find the midpoint J of OB and then the midpoint I of OJ. Then the length of OI is $\frac{1}{4}$.
4. Bisect the angle OIP_1 and then bisect the angle between OI and the bisector of OIP_1 to get a line through I meeting OP_1 at E. Then the angle OIE is one quarter of the angle OIP_1.
5. Draw the line through I at right-angles to IE and bisect the angle between this line and IE to get a line through I meeting P_1O produced at F. Then the angle EIF is 45 degrees.
6. Draw the circle with P_1F as diameter, by first constructing its centre which is the midpoint of P_1F (this point is not marked on Fig. 20 as it is too close to other marked points). Let K be the point where this circle meets OB.
7. Draw the circle centre E through K. Let it meet OP_1 at N and P_1O produced at M.
8. Draw lines through N and M at right-angles to OP_1 meeting the original circle centre O through P_1 at P_4 and P_6.
9. Draw the circle centre P_4 through P_6 and let P_2 be the other point where it meets the original circle centre O through P_1.

10. Then the angle P_1OP_2 is $360/17$ degrees, so the remaining vertices of the regular 17-gon can be found by marking successive points around the circumference of the original circle such that the distance between each adjacent pair is the distance between P_1 and P_2.

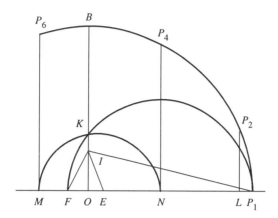

Fig. 20. Construction of a regular 17-gon

Recall that the constructibility of the 17-gon is equivalent to the constructibility of the length of OL, where L is the foot of the perpendicular from P_2 onto OP_1. Since the length of OL is a constructible number we know that it must be expressible in terms of the integers and square roots, addition, subtraction, multiplication and division (see Theorem 3). In fact it is equal to

$$\frac{-1 + \sqrt{17} + \sqrt{34 - 2\sqrt{17}} + \sqrt{68 + 12\sqrt{17} - 16\sqrt{34 + 2\sqrt{17}} - 2(1 - \sqrt{17})\sqrt{34 - 2\sqrt{17}}}}{16}$$

(see p. 173 of Stewart (1989)).

Solutions to Worksheet 1

1. The distances between the two points P_1 and P_2 first constructed on the circumference of the circle and between them and the centre O are all equal to the radius of the circle, so OP_1P_2 is an equilateral triangle (see Fig. 21). Therefore its angles are all the same and so, since the angles of any triangle add up to 180 degrees, the angles of OP_1P_2 are all 60 degrees. As $60 = 360/6$, this shows that the construction works.

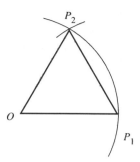

Fig. 21

2. The lengths of the corresponding sides of the triangles OAC and OBC are all the same (see Fig. 22), so the triangles are congruent (that is, they have the same size and shape).

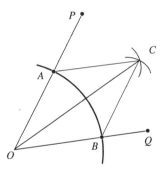

Fig. 22.   Bisecting an angle

3. By symmetry (consider Fig. 23). Alternatively the triangles PAB and QAB are congruent (because their corresponding sides have the same lengths), so if D is the point where AB meets PQ then the triangles PAD and QAD are

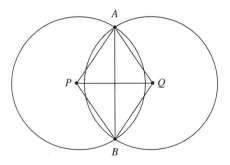

Fig. 23.   Construction of a perpendicular bisector

congruent (because they have two pairs of corresponding sides with the same lengths, and the included angles are equal).

For the last part, find the point R on the line PQ extended beyond P which is the same distance from P as Q is, and construct the perpendicular bisector of QR (see Fig. 24).

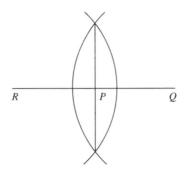

Fig. 24. Construction of the line meeting PQ at right-angles at P

4. The perpendicular bisector of the line segment PQ will pass through R and meet the line L at right-angles.
5. First measure distance x along the given line segment to construct the point $(x, 0)$. Now use Exercise 3 to construct the line through $(x, 0)$ at right angles to the x-axis and measure distance y along this to find (x, y). Conversely given (x, y) drop a perpendicular from it to the x-axis using Exercise 4; this perpendicular will meet the x-axis at the point $(x, 0)$, and the line segments joining the origin to this point and this point to (x, y) will have lengths x and y.
6. We have

$$a\left(x + \frac{b}{2a}\right)^2 = a\left(x^2 + \frac{bx}{a} + \frac{b^2}{4a^2}\right)$$

which equals

$$ax^2 + bx + c - c + \frac{b^2}{4a} = \frac{b^2}{4a} - \frac{4ac}{4a}.$$

Hence

$$x + \frac{b}{2a} = \pm\sqrt{\frac{b^2 - 4ac}{4a^2}} = \pm\frac{\sqrt{b^2 - 4ac}}{2a}.$$

7. For the first part, the triangles with vertices (x, y), (a, b), (x, b) and (c, d), (a, b), (c, b) are similar, so

$$\frac{x - a}{c - a} = \frac{y - b}{d - b}.$$

The second part follows straight from Pythagoras' theorem.

8. By Exercise 7 $(c-a)(y-b)=(d-b)(x-a)$, so

$$(c-a)(y-f) = (c-a)(y-b)+(c-a)(b-f)$$
$$= (d-b)(x-a)+(c-a)(b-f).$$

Now multiply the second equation of Exercise 7 through by $(c-a)^2$ and substitute for $(c-a)(y-f)$ to obtain the required quadratic equation in x. Note that the coefficient of x^2 is the sum of two non-negative numbers which cannot both be zero, so it is nonzero. Now apply Exercise 6.

9. Subtract the equation for one circle from that of the other, to obtain the linear equation

$$2(h-e)x + 2(k-f)y = r^2 - s^2 + h^2 - e^2 + k^2 - f^2.$$

Now multiply the equation of the first circle through by $4(k-f)^2$ and substitute for $2(k-f)y$ in it using the equation above to get a quadratic equation satisfied by x. Now proceed as in Exercise 8.

Solutions to Worksheet 2

1. If $n = mk$, join up every kth vertex of the regular n-gon.
2. (i) We need to construct the centre of the circle. If n is even it is the intersection of the line segments joining pairs of opposite vertices, whereas if n is odd it is the intersection of the line segments joining vertices to the midpoints of the opposite edges.

   (ii) The angle between the line segments joining the centre of the circle in (i) to any two adjacent vertices is $360/n$ degrees.

   Conversely given two line segments meeting at a point O at an angle of $360/n$ degrees, construct a circle centre O small enough to meet both the line segments, at points P_1 and P_2, say. Then spread the compasses to the distance between P_1 and P_2 and use them to mark points the same distance apart all the way around the circumference of the circle. These points will be the vertices of a regular n-gon.

3. By Exercise 2 we can construct an angle of $360/n$ degrees. By Exercise 2 of Worksheet 1 we can bisect this angle to get an angle of $360/2n$ degrees, and by Exercise 2 again this means we can construct a regular $2n$-gon.

   For a 12-gon we can construct the first two vertices of a regular hexagon inscribed in a circle as in Example 1, bisect the angle between the line segments joining the two vertices to the centre of the circle as in Exercise 2 of Worksheet 1 and thus obtain the first three vertices of a regular 12-gon inscribed in the circle. We then measure off the remaining vertices round the circumference.

4. We have

$$57 = (2 \times 25) + 7$$
$$25 = (3 \times 7) + 4$$
$$7 = (1 \times 4) + 3$$
$$4 = (1 \times 3) + 1$$

so that

$$1 = 4 - (1 \times 3)$$
$$= 4 - (1 \times (7 - 1 \times 4))$$
$$= (2 \times 4) - (1 \times 7)$$
$$= (2 \times (25 - 3 \times 7)) - (1 \times 7)$$
$$= (2 \times 25) - (7 \times 7)$$
$$= (2 \times 25) - (7 \times (57 - 2 \times 25))$$
$$= (16 \times 25) - (7 \times 57).$$

5. By constructing angles adjacent to each other we can construct the sum of any two constructible angles. By constructing angles inside each other we can construct differences. Repeating these constructions enough times will give us what we want.

6. If $an + bm = 1$ then $a(360/m) + b(360/n) = 360/nm$ (multiplying both sides by $360/nm$). Therefore by Exercises 2 and 5 we can construct an angle of $360/nm$ degrees, and thus a regular nm-gon.

Figure 18 contains two adjacent vertices of a regular 15-gon inscribed in the circle.

7. The first three Fermat primes are 3, 5 and 17. The fourth and fifth are 257 and 65 537. There are no more Fermat primes less than one million since

$$2^{2^5} = 2^{32} = (2^4)^8 = 16^8 > 10^6.$$

(In fact there are no more Fermat primes known: 2^{2^r} is not prime for all the values of r greater than or equal to 5 for which it has been tested).

8. The numbers n less than or equal to 100 such that a regular n-gon is constructible with straight-edge and compasses are 3, 4, 5, 6, 8, 10, 12, 15, 16, 17, 20, 24, 30, 32, 34, 40, 48, 51, 60, 64, 68, 80, 85, 96.

Acknowledgements

My main source for the material in this masterclass was Ian Stewart's book on Galois theory (1989). In particular Fig. 20 is a slight modification of Fig. 27 in Stewart (1989).

I am very grateful to all those who organized and/or took part in those series of masterclasses at the University of Reading in which I participated, for providing both stimulus and feedback; I found the masterclasses extremely enjoyable experiences.

References and Further Reading

J. Baylis and R. Haggarty, *Alice in Numberland* (Macmillan, London, 1988)

C. B. Boyer, *A History of Mathematics* (Wiley, New York, 1968)

R. Courant and H. Robbins, *What is Mathematics?* (Oxford University Press, 1941)

H. S. M. Coxeter, *Introduction to Geometry* (Wiley, New York, 1969)

D. J. H. Garling, *A Course in Galois theory* (Cambridge University Press, 1986)

I. Stewart, *The Problems of Mathematics* (Oxford University Press, 1987)

I. Stewart, *Galois theory* (2nd edn) (Chapman and Hall, London, 1989)

6 Discrete Mathematics and its Application to Ecology

by PHILIP MAINI

Centre for Mathematical Biology, Mathematical Institute, University of Oxford

1. Introduction

In this chapter we will illustrate how simple mathematical ideas can be used to understand how a population grows. We first consider the well-known problem which generates the *Fibonacci sequence*. Consider a population of rabbits. Suppose that every pair of rabbits can reproduce only twice, when they are one- and two-months old, and that each time they produce exactly one new pair of rabbits. Assume that all newborn rabbits survive. The problem here is to see if we can predict how many newborn rabbits there will be in a certain generation given an initial newborn rabbit population.

Before beginning, we need to use a convenient notation. Let us define the quantity N_m to be the number of newborn pairs in generation m. Suppose that we begin with one newborn pair. Then we say that N_0, the number of newborn pairs in the zero generation (the Adam and Eve of rabbits), is one. Now, after one month, this pair of rabbits gives birth to one pair of rabbits. Therefore, we have that N_1, the number of newborn pairs in the first generation, is one. At the end of the second month, Adam and Eve give birth to their second newborn pair and then will no longer give birth to any more rabbits. But, the pair of rabbits in the first generation also give birth to a pair of newborns (Adam and Eve's 'grandbunnies'). Therefore, N_2, the number of newborn pairs in the second generation, will be $1 + 1 = 2$. In the third generation, newborn pairs will arise from those pairs that were newborns in the second generation (N_2), and from those newborns in the first generation (N_1). Therefore the number of newborns in the third generation, N_3, will be $N_2 + N_1 = 2 + 1 = 3$. The number in the fourth generation, N_4, will be $N_3 + N_2 = 3 + 2 = 5$ (see Fig. 1). We can now prove the following theorem.

Theorem 1 (The Rabbit Problem)
If

$$N_m = \text{number of newborn pairs in generation } m$$

then

$$N_{m+1} = N_m + N_{m-1}. \tag{1}$$

Proof
The number of newborn rabbits in the $(m + 1)$th generation will come either from one-month old parents who were born in the mth generation or from two-month old parents who were born in the $(m - 1)$th generation (see Fig. 1). Hence result.   □

To see how this ties in with the above discussion, let us suppose that $N_0 = 1$ (this means that we begin with one pair of newborns in the zero generation). Then, we know that $N_1 = 1$. Now, if we put $m = 1$ into equation (1), we have that $N_2 = N_1 + N_0$, that is, $N_2 = 1 + 1 = 2$, which is the number of newborns in the second generation. Now putting $m = 2$ into equation (1), we have that $N_3 = N_2 + N_1 = 2 + 1 = 3$, the number of newborns in the third generation. For $m = 3$, we have $N_4 = 3 + 2 = 5$, the number of newborns in the fourth generation. Continuing like this we have that $N_5 = 5 + 3 = 8$, $N_6 = 8 + 5 = 13$, $N_7 = 13 + 8 = 21$, etc.

The sequence of numbers $1, 1, 2, 3, 5, 8, 13, 21, \ldots$ is known as the *Fibonacci sequence*. Equation (1) is an example of a *recurrence relation* and this type of relation is very common in ecological studies. A mathematical formulation of such a problem is known as a *mathematical model* of the problem. The numbers generated by the Fibonacci sequence are known as the *Fibonacci numbers* and occur widely in nature. For example, the numbers of spirals running in opposite directions on a ripening sunflower or a pine cone are often consecutive Fibonacci numbers.

We now move on to what appears, to begin with, to be a very different problem. We call it the *Beetle Problem*. Suppose that beetles of one generation produce the larvae (immature form of an insect which eventually becomes the insect) of the next generation and then die so that only the larvae survive. Assume that all the larvae survive. The problem here is, can we predict the number of larvae in a certain generation? In this case, we define N_m to be the number (in some units) of larvae in generation m. Suppose that the average

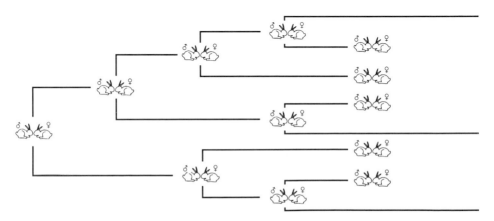

Fig. 1.   Proof of Theorem 1 (see text for details)

number of larvae produced per beetle is 2. Let us assume that in generation 0, N_0 is 1. It is important to note here that setting N_1 to be one does not necessarily mean that in the first generation there is one beetle. This is because we have not specified our units. For example, if the unit was thousands, then $N_0 = 1$ would mean that there were one thousand beetles in generation 0. Note that $N_0 = \frac{1}{2}$ would not mean half a beetle, but would mean 500 (one half of a thousand) beetles.

Assuming that $N_0 = 1$, we have that N_1, the number of beetles in the first generation, is $2 \times 1 = 2$. The beetles in the second generation arise from the larvae produced by the first generation, so N_2 is $2 \times N_1 = 4$. In the same way, $N_3 = 2 \times 4 = 8$ etc. In fact, we can now prove the following theorem.

Theorem 2 (The Beetle Problem)
If

$$N_m = \text{number of beetles (in some units) in generation } m$$

$$r = \text{average number of larvae produced per beetle}$$

then

$$N_m = N_0 r^m \qquad (2)$$

Proof
The generation m beetles come from the $(m-1)$th generation. Now each beetle in the $(m-1)$th generation produced r larvae, so $N_m = rN_{m-1}$. Applying the same argument to the $(m-1)$th generation, we have $N_{m-1} = rN_{m-2}$. So we can write $N_m = r^2 N_{m-2}$. If we carry on this reasoning we deduce that $N_m = N_0 r^m$ (see Fig. 2).   □

For our example, $r = 2$ and $N_0 = 1$. The formula then says that $N_1 = N_0 \times 2 = 2$, $N_2 = N_0 \times 4 = 4$, $N_3 = N_0 \times 8 = 8$, etc.

Notice that if we want to calculate the number of beetles in a certain generation, then all we need to know is N_0 and r and we have immediately that $N_m = N_0 r^m$. However, for the rabbit problem, to find N_m, we need to know N_{m-1} and N_{m-2}. But to calculate these values, we need to know N_{m-3}, etc. Therefore, if m is very large, then the calculations involved for the beetle problem are easier than those for the rabbit problem. However, in question 6 of Worksheet 1 we will see that there is an easier way to calculate N_m for the rabbit problem, when m is large.

1. Consider the Fibonacci numbers $1, 1, 2, 3, 5, 8, 13, 21, 34, 55, 89, 144, 233, \ldots$
   (a) Calculate (to 3 decimal places) the fractions obtained from dividing each number by the previous number, that is, $\frac{1}{1}, \frac{2}{1}, \frac{3}{2}, \frac{5}{3}, \frac{8}{5}, \frac{13}{8}, \frac{21}{13}, \frac{34}{21}, \frac{55}{34}, \frac{89}{55}, \frac{144}{89}, \frac{233}{144}$.
   (b) What do you notice about the values you calculated in (a)?
   (c) The calculations in (a) and (b) suggest that for two consecutive numbers N_{m+1}, N_m, in the Fibonacci sequence, $\dfrac{N_{m+1}}{N_m} \approx r$ for large m. What is that value of r?

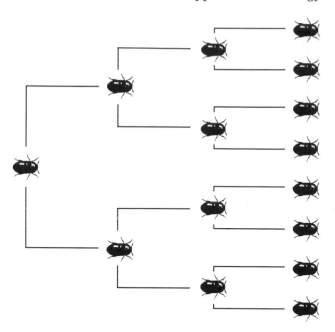

Fig. 2. The Beetle problem for the case $r = 2$. Note the differences between this figure and Fig. 1

(d) Is the Rabbit problem *really* that different from the Beetle problem?
2. From equation (2) calculate N_1, N_2, N_3 and N_4 in the following cases.
    (a) $N_0 = 2.0$, $r = 2.0$.
    (b) $N_0 = 2.0$, $r = 0.5$.
3. From equation (2) what size will N_m be for m large in the following cases.
    (a) $r > 1.0$.
    (b) $r = 1.0$.
    (c) $r < 1.0$.
    (d) Do you think that equations (1) and (2) are good mathematical models?

For the following exercises assume that the recurrence relation has the form

$$N_{m+1} = rN_m(1.0 - N_m). \tag{3}$$

(Note that if N_m is small, equation (3) looks a bit like equation (2)).

4. Calculate N_1, N_2, N_3 and N_4 in the following cases.
    (a) $N_0 = 0.6$, $r = 0.5$.
    (b) $N_0 = 0.4$, $r = 2.0$.
    (c) $N_0 = 0.6$, $r = 2.0$.
    (d) $N_0 = 0.5$, $r = 2.0$.
    (e) In (b), (c) and (d) what do you think the value of N_m would be for m very large?
    (f) Is this a better mathematical model than equations (1) and (2)?

5. Calculate N_1, N_2, N_3, N_4 and N_5 (each to 2 decimal places) for equation (3) in the case where $r = 3.1$, $N_0 = 0.70$. What do you think are the values of $N_6, N_7, N_8, N_9, \ldots$?

6. In question 1, we made an approximate calculation for r. We can make a more exact calculation as follows. Suppose that $\dfrac{N_{m+1}}{N_m} = \dfrac{N_m}{N_{m-1}} = r$.

   (a) From equation (1) show that $N_{m+1} = \left(1 + \dfrac{1}{r}\right)N_m$.

   (b) From (a), show that $1 + \dfrac{1}{r} = r$.

   (c) From (b), calculate r.

2. The logistic map

In Worksheet 1, question 3, we saw that the models presented in Section 1 have their limitations, and we introduced a new model

$$N_{m+1} = rN_m(1.0 - N_m). \tag{4}$$

This model is called the *logistic model*, or the *logistic map*. Before we analyse this model further we need the following definition.

Definition
If the successive values calculated from a recurrence relation tend towards some fixed value, we call this value the *steady state value*. It has the property that $N_{m+1} = N_m = N^*$, the steady state value, for large m.

The steady states for equation (4) are given by

$$N^* = 0 \quad \text{or} \quad N^* = \frac{r-1}{r}. \tag{5}$$

For $0 < r < 1$ the second steady state does not make sense because it is negative and if we start off with $0 \leq N_0 \leq 1$ in equation (4) then $0 \leq N_m \leq 1$ for every m. So the only sensible solution is the first solution of equation (5) (compare this with Worksheet 1, question 4(a)). For $1 < r < 3$ the second solution makes sense and it appears from Worksheet 1, question 4(e), that this is the steady state, at least for $r = 2.0$. Hence, although there are now two possibilities for the steady state, we have moved off the first solution onto the second. This is called a *bifurcation*. For $r = 3.1$, the second solution of equation (5) is still a sensible steady state but Worksheet 1, question 5 suggests that we have a 2-*cycle solution*. We may summarize these results on the *bifurcation diagram* in Fig. 3. Note that we have been careful not to go too far beyond $r = 3$ in case there are yet more surprises in store!

We can get an idea of what the solution to equation (4) will look like by using a graphical technique called *cobwebbing*. The technique is illustrated in Fig. 4 and the results represented graphically in Fig. 5.

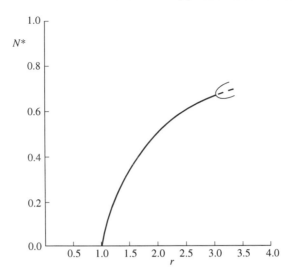

Fig. 3. Bifurcation diagram for equation (4) showing how the steady state N^* varies with r. The dashed lines are *unstable solutions*

We draw the curve of equation (4) (with $r = 0.5$) and also the straight line $N_{m+1} = N_m$. We first find the value 0.6 along the horizontal axis. This is N_0. Now $N_1 = 0.5 \times N_0(1 - N_0)$. But this is exactly the length of the vertical straight line from the point N_0 to the curve. If we then draw from there a line parallel to the horizontal axis, then the distance this line is above the horizontal axis is always

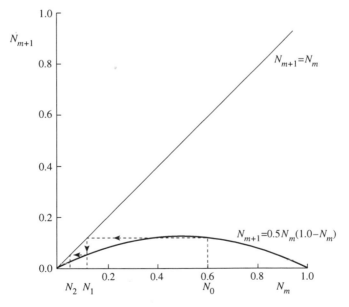

Fig. 4. Illustration of the cobwebbing technique for equation (4) with $N_0 = 0.6$, $r = 0.5$

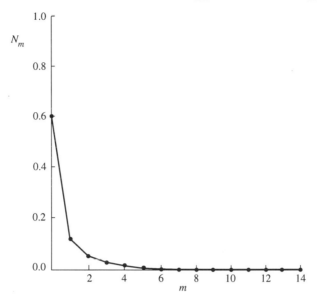

Fig. 5. Graphical representation of the solution obtained by the cobwebbing technique illustrated in Fig. 4 (Compare this solution with that from Worksheet 1, question 4)

N_1. If we drop a perpendicular line onto the horizontal axis from the point where this line intersects the $N_{m+1} = N_m$ line, we have that the distance from this point to the origin is also N_1. That is, given N_0, we have been able to calculate, graphically, the value of N_1. In a similar way we can now go on to calculate $N_2, N_3, \ldots$, etc.

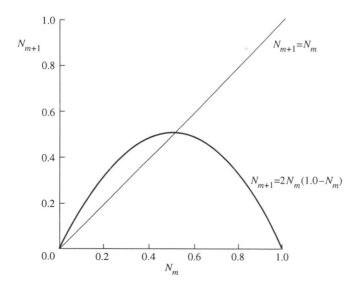

Fig. 6. Cobweb for Worksheet 2, question 1

Worksheet 2

1. Use Fig. 6 to cobweb the solution to equation (4) for the case $N_0 = 0.7$, $r = 2.0$.
2. Illustrate your results to question 1 graphically. Does this solution cycle or does it go to a steady state?
3. Use your calculator to find N_1, N_2, N_3, N_4, N_5, N_6, N_7, and N_8 (each to 3 decimal places) for equation (4) in the case $N_0 = 0.501$, $r = 3.5$. Draw the cobweb diagram on the following figure (Fig. 7).

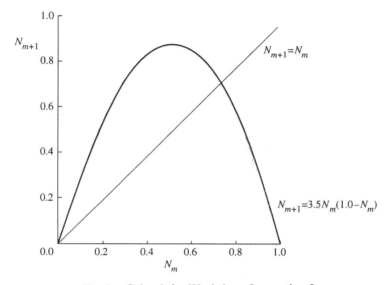

Fig. 7. Cobweb for Worksheet 2, question 3

4. Illustrate your results to question 3 graphically.
5. From equation (4), what is the equation for N_{m+2} in terms of N_m? What are the steady states of this equation?

3. Chaos

From Worksheet 2, question 3, we saw that if $r = 3.5$, then equation (4) has a *4-cycle* solution. As r increases further, we get an 8-cycle solution and then a 16-cycle solution, etc. We can represent these results (Fig. 8) on the bifurcation diagram we began to draw in Fig. 3.

If we increase r beyond a critical value $r_c \approx 3.83$ we lose the periodic structure of the solution. Figure 9 illustrates the case when $r = 3.9$. In this case the solution does not have a periodic structure. It is said to be *chaotic*. If we were to change the value of N_0 slightly we would get a completely different looking solution. Another chapter in this book, 'A Little Bit of Chaos', introduces you to other aspects of chaos.

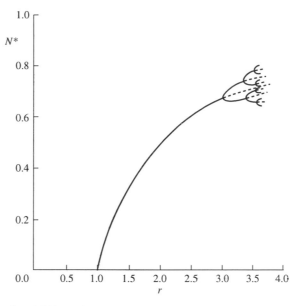

Fig. 8. The completed bifurcation diagram illustrating, schematically, the *cascade of period doubling bifurcations*

Equation (4) is a good model for insect growth – many insect populations either reach a steady state or oscillate in a 2-cycle (see Fig. 10).

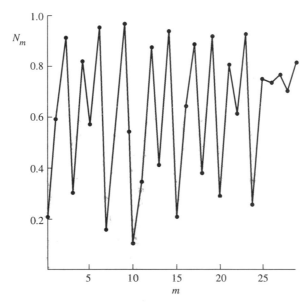

Fig. 9. The solution to equation (4) for the case $r = 3.9$, $N_0 = 0.2$. Note that for some generations the population gets very large then crashes back in the following generation

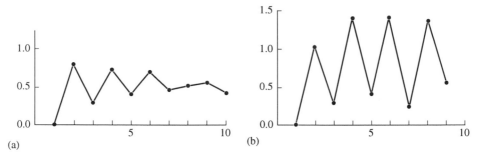

Fig. 10. Examples of population growth of two types of beetles. The vertical axis is the number of beetles (thousands), the horizontal axis is the generation number. (Redrawn from Hassel (1976))

Note that for $r > 4$ the model predicts negative populations and thus is not realistic in this case. You may want to do a few iterations or draw some cobweb maps to see why (hint: start off with $r = 4.1$, $N_0 = 0.5$).

The analysis of the model equation (4) suggests that if we let r get too large then we will get cycling populations with very large numbers or chaotic populations with the possibility of very large numbers. Therefore, in order to control the population numbers, we really need to keep $r < 3.83$ and preferably much smaller. We can decrease r by, for example, releasing sterile insects into the population, so that the average birth rate per insect decreases.

We can write down more complicated relations, or *iterated maps*. For example, we could have

$$N_{m+1} = \frac{rN_m}{(1.0 + aN_m)^b} \tag{6}$$

where r, a and b are positive constants. Figure 11 shows the results from equation (6) for the case $r = 20$, $a = 0.01$, $N_0 = 0.01$ and different values of b.

4. Summary

In this chapter we have illustrated how mathematical models can be used to study population growth. The analysis of these models determines if the assumptions we made in forming the model are sensible. If not, we have to reformulate the model. This helps us to understand more deeply the processes of population growth that we are trying to model and also predicts how the population will react if we carry out certain experiments. The role of a predictive mathematical model is very important. For example, it can aid our understanding of how certain diseases spread within a population and help us to choose the best control strategy. Mathematical models are important not only in ecology, but also in the fields of biology, physics, chemistry and medicine.

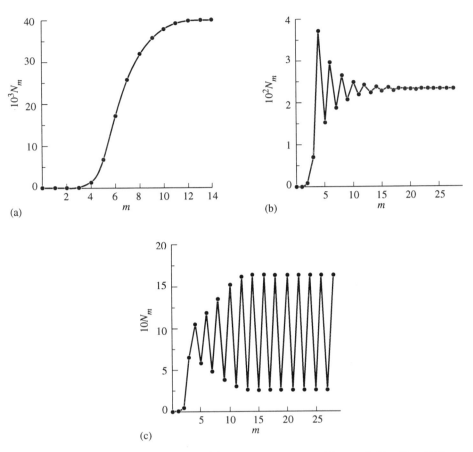

Fig. 11. The solution to equation (6) for the case $r = 20$, $a = 0.01$, $N_0 = 0.01$ and different values of b: (a) $b = 0.5$, (b) $b = 2.5$, (c) $b = 5.0$. The solution shown in (a) is very similar to the behaviour of growing bacteria and of yeast

Solutions to Worksheet 1

1. (a) 1, 2, 1.500, 1.667, 1.600, 1.625, 1.615, 1.619, 1.618, 1.618, 1.618, 1.618.
   (b) The values calculated in (a) tend to 1.618 to 3 decimal places. (In fact, this is the golden mean $\dfrac{1 + \sqrt{5}}{2}$).
   (c) $\dfrac{N_{m+1}}{N_m} \approx r$, where $r \approx 1.618$ to 3 decimal places.
   (d) The Rabbit problem, for large m, is a special case of the Beetle problem with $r = 1.618$ to 3 decimal places.
2. (a) 4.0, 8.0, 16.0, 32.0.
   (b) 1.0, 0.5, 0.25, 0.125.
3. (a) N_m will be very large.
   (b) $N_m = N_0$.

(c) N_m will be very small.

(d) Equation (1) predicts that the population of rabbits will grow forever; equation (2) predicts that the population of beetles will either grow and grow (if $r > 1.0$), or die down to zero (if $r < 1.0$) or stay at its initial value (if $r = 1.0$). Neither of these models is very good because populations do not behave in this way. For example, death has been ignored.

4. (a) 0.120, 0.053, 0.025, 0.012 to 3 decimal places.

(b) 0.480, 0.499, 0.500, 0.500 to 3 decimal places.

(c) 0.480, 0.499, 0.500, 0.500 to 3 decimal places.

(d) 0.500, 0.500, 0.500, 0.500.

(e) N_m tends to 0.5 for large m.

(f) This seems a better model because for certain r-values the population reaches a steady value.

5. 0.65, 0.71, 0.65, 0.71, 0.65, 0.71, 0.65, 0.71,...

6. (a) From the assumption, $N_{m-1} = N_m/r$, therefore equation (1) becomes $N_{m+1} = N_m + N_m/r$.

(b) From (a), we have $N_{m+1} = (1 + 1/r)N_m$. The assumption says that $N_{m+1} = rN_m$. Therefore $1 + 1/r = r$.

(c) From (b), $r^2 - r - 1 = 0$. Therefore $2r = 1 \pm \sqrt{5}$. The negative solution does not make sense because r is assumed positive, therefore $r = (1 + \sqrt{5})/2$.

Solutions to Worksheet 2

1.

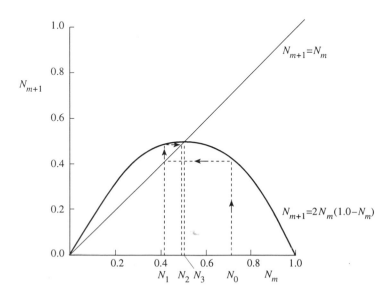

Fig. 12 Cobweb diagram for Worksheet 2, question 1

2.

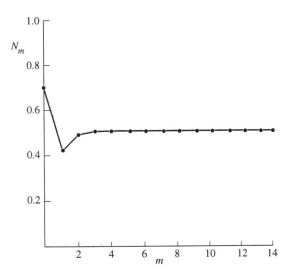

Fig. 13 Graphical representation of the solution obtained by the cobwebbing technique for Worksheet 2, question 1. The solution goes to a steady state

3. 0.875, 0.383, 0.827, 0.501, 0.875, 0.383, 0.827, 0.501

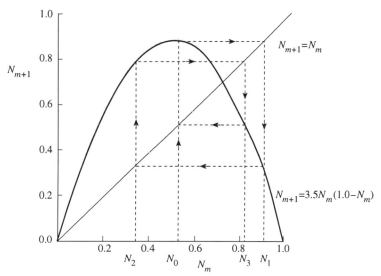

Fig 14 Cobweb diagram for Worksheet 2, question 3

4.

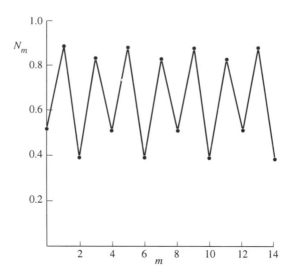

Fig 15 Graphical representation of the solution obtained by the cobwebbing technique for Worksheet 2, question 3

5. From equation (4) we can write $N_{m+2} = rN_{m+1}(1.0 - N_{m+1})$. Using equation (3) again, we have $N_{m+1} = rN_m(1.0 - N_m)$, so we can write

$$N_{m+2} = r^2 N_m(1.0 - N_m)(1.0 - rN_m(1.0 - N_m)). \tag{7}$$

The steady states of equation (7) are given by $N_{m+2} = N_m = N^*$. Thus we have

$$N^* = r^2 N^*(1.0 - N^*)(1.0 - rN^*(1.0 - N^*)) \tag{8}$$

which has solution $N^* = 0$ or

$$(N^*)^3 - 2(N^*)^2 + \left(1 + \frac{1}{r}\right)N^* + \frac{1}{r^3} - \frac{1}{r} = 0. \tag{9}$$

We know that $1 - \frac{1}{r}$ is also a steady state so we can factor this out from equation (9) to get

$$\left(N^* - 1 + \frac{1}{r}\right)\left((N^*)^2 - \left(1 + \frac{1}{r}\right)N^* + \frac{1}{r} + \frac{1}{r^2}\right) = 0. \tag{10}$$

Thus the two further solutions are given by the roots of $((N^*)^2 - (1 + 1/r)N^* + 1/r - 1/r^3) = 0$, that is, $2N^* = 1 + 1/r \pm [(1 + 1/r)^2 - (4/r)(1 + 1/r)]^{1/2}$. This may be written as $2N^* = 1 + (1/r) \pm (1/r)[(r - 1)^2 - 4]^{1/2}$. Note that this solution will make sense only if $r > 3$, so we have a 2-cycle solution possible if $r > 3$.

References

R. Anderson and R. M. May, 'The logic of vaccination,' *New Scientist* (18th November, 1982), 410–15

O. M. Hassel, *The dynamics of competition and predation* (Edward Arnold Publishers, 1976)

R. M. May (editor), *Theoretical Ecology, Principles and Applications* (2nd edition) (Blackwell Scientific Publications, 1981)

7 Channels, Pipes and Rivers

by DAVID NEEDHAM

Department of Mathematics, University of Reading

1. Observations and examples

The flow of water arises in many everyday situations. For example, on a large scale, in the oceans, in rivers and canals and in lakes, or, on a smaller scale, in heating or drainage pipes and channels. In this class we will develop a simple *mathematical model* to describe and characterize water flows which occur in closed and open channels. Examples of such flows include the following:

(a) Closed channels

Central heating pipes, drainage pipes, closed aqueducts, reservoir outlets, liquid fuel pipes, fire hoses.

(b) Open channels

Rivers, canals, irrigation channels, spillways from reservoirs and lakes, open drainage channels, guttering.

We will first consider some basic observations concerning the mechanisms involved in the flow of water.

2. The flow of water

Suppose that we observe a body of water flowing from left to right, as shown in Fig. 1.

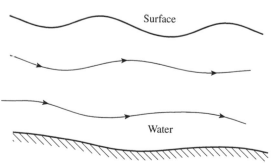

Fig. 1. Vertical section of river

Figure 1 gives a vertical cross section through a river flowing from left to right. We may ask the following questions.

(a) What is the mechanism which makes the water flow?
(b) How can we describe the flow?

The answer to (a) is that neighbouring elements (or blobs) of water exert *forces* on each other, which cause their motion. To characterize these internal forces acting on water elements we introduce water *pressure*, which measures (at each point in the body of water) the *force per unit area* acting on water elements. We denote the water pressure by p (whose value will depend on the position in the water at which we are measuring). An illustration is shown in Fig. 2.

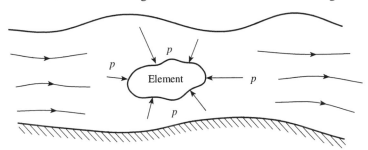

Fig. 2. Pressure acting on an element of water

In Fig. 2, the length of the arrows represent the magnitude of the pressure, and, in general, this will vary from position to position on the boundary of the water element.

The answer to (b) is that the water flow is fully described once we know the *water pressure*, p, and the *water speed*, v, at each point in the region occupied by the water.

Dimensions

In S.I. units

> water speed v is measured in   *metres per second* (ms^{-1})

> water pressure p is measured in   *Newtons per square metre* (Nm^{-2})

Note that atmospheric pressure, $P_A \simeq 10^5 \times 1.012\,\text{Pascal}$, where $1\,\mathrm{Nm}^{-2}$ is called a Pascal.

Variables

v and p are the *variables* (their value depends upon position in the flow and time) which describe the water flow.

Incompressibility

An important physical property of water is that it is *incompressible*. This means that a fixed mass of water always occupies the same volume, no matter what the pressure. This is easily verified by filling a (strong) tube half full of water and inserting a well-fitting plunger into the open end. However much force is exerted onto the plunger, it will not descend below the original water level. The water cannot be 'squashed' into a smaller volume by applying pressure (see Fig. 3).

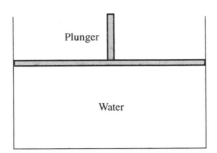

Fig. 3. Incompressibility of water

Since water is incompressible, the *density* (mass per unit volume) of water is *constant* at all points in the flow. We denote this by ρ, which has dimensions kilograms per cubic metre ($kg\,m^{-3}$). The density of water will be taken as $10^3\,kg\,m^{-3}$.

3. Mathematical model for flow in closed channels

Consider water flowing steadily (so that conditions do not change in time) through a closed horizontal channel of rectangular cross-section and variable vertical width, as shown in Fig. 4.

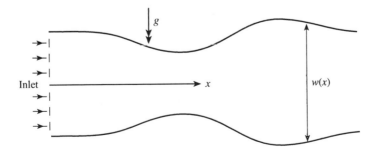

Fig. 4. Vertical cross-section along the horizontal channel

Let x be a *coordinate* measuring horizontal distance along the channel from the inlet (at $x = 0$). The vertical width of the channel w varies as we move along, and so w is a function of x. We write $w = w(x)$. The horizontal breadth of the channel does not change with x and is denoted by b_0. A cross-section of the channel is shown in Fig. 5.

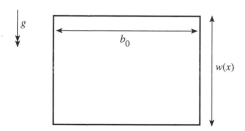

Fig. 5.   Vertical cross-section showing the breadth of the channel

The centre line of the channel lies *horizontally* whilst the *force of gravity* acts vertically downwards. The constant g is the *acceleration due to gravity* and $g \approx 10 \, \mathrm{ms}^{-2}$ in S.I. units.

Water enters the inlet of channel which has width w_0, with speed v_0 and flows horizontally through the channel (from left to right in Fig. 4).

The *volumetric flow* $Q(x)$ across the cross-section of the channel at a distance x from the inlet is the total volume of water (m^3) passing through this cross-section in one second. $Q(x)$ has dimensions $\mathrm{m}^3\mathrm{s}^{-1}$.

At $x = 0$,

$Q_0 = Q(0) =$ water speed at ($x = 0$) $\times$ channel cross-sectional area at ($x = 0$)

$\qquad = v_0 \times (w_0 \times b_0)$.

Therefore, $\qquad\qquad\qquad\qquad Q_0 = v_0 w_0 b_0.$ (1)

At a distance x from the inlet,

$\qquad Q(x) =$ water speed at $x \times$ channel cross-sectional area at x

$\qquad\qquad = v(x) \times (w(x) \times b_0)$.

Therefore, $\qquad\qquad\qquad\qquad Q(x) = v(x)w(x)b_0.$ (2)

However, water is *incompressible*, so the volume of water entering at the inlet over one second must be the same as the volume of water leaving through the cross-section at x over one second. This requires

$\qquad\qquad\qquad\qquad Q(x) = Q_0$ at each cross-section x.   (3)

Substituting from equations (1) and (2) into (3) gives

$$v(x)w(x)b_0 = v_0 w_0 b_0.$$

On cancelling b_0, this becomes

$$v(x)w(x) = v_0 w_0 \text{ at each } x. \qquad\qquad (4)$$

This is called an equation of *conservation of mass* for the flow of water in the channel. Once we know $w(x)$, then equation (4) determines the water speed $v(x)$ at any cross-section x along the pipe. Dividing through equation (4) by $w(x)$ we find

$$v(x) = \frac{v_0 w_0}{w(x)} \text{ at each } x. \tag{5}$$

A further equation relating water pressure $p(x)$ and speed $v(x)$ is obtained by application of *Newton's second law of motion* to the water in the channel. This requires that the quantity

$$p(x) + \tfrac{1}{2}\rho v(x)^2, \tag{6}$$

remains the same (constant) at any cross-section x in the channel. When the pressure at the inlet is p_0, then equation (6) becomes the *Bernoulli equation*

$$p(x) + \tfrac{1}{2}\rho v(x)^2 - p_0 + \tfrac{1}{2}\rho v_0^2 \text{ at each } x. \tag{7}$$

To obtain an explicit equation for $p(x)$, we may now substitute from equation (5) into equation (7). This gives

$$p(x) + \tfrac{1}{2}\rho \frac{(v_0 w_0)^2}{w(x)^2} = p_0 + \tfrac{1}{2}\rho v_0^2 \text{ at each } x,$$

which rearranges to

$$p(x) = p_0 + \tfrac{1}{2}\rho v_0^2 \left(1 - \frac{w_0^2}{w(x)^2}\right) \text{ at each } x. \tag{8}$$

Equations (5) and (8) determine the water speed $v(x)$ and water pressure $p(x)$ at each cross-section x down the channel from the inlet at $x = 0$. These are determined in terms of the condition at the inlet p_0, v_0, w_0 and the width of the channel $w(x)$.

Pressure

(a) When $w(x) = w_0$, then $p(x) = p_0$.
(b) When $w(x) > w_0$, then $p(x) > p_0$.
(c) When $w(x) < w_0$, then $p(x) < p_0$.

Speed

(a) When $w(x) = w_0$, then $v(x) = v_0$.
(b) When $w(x) > w_0$, then $v(x) < v_0$.
(c) When $w(x) < w_0$, then $v(x) > v_0$.

We have demonstrated the following.

Theorem 1

In a horizontal channel of rectangular cross-section, the water speed *increases* at *constrictions* and *decreases* at *expansions*. Correspondingly, the water pressure *decreases* at *constrictions* and *increases* at *expansions*.

Example 1

A uniform channel starts at $x = 0$ and has width w_0 which is constant (see Fig. 6).

Fig. 6.   Uniform channel

To find $p(x)$ and $v(x)$ we return to equations (5) and (8). We substitute for $w(x)$ with w_0 to obtain

$$\left. \begin{array}{l} p(x) = p_0 \\ v(x) = v_0 \end{array} \right\} \text{ at each cross-section } x.$$

The pressure and speed remain the same at each cross-section along the channel and are equal to their values at the inlet.

Example 2

An expanding channel has width w_0 at $x = 0$, whilst the width changes to

$$w(x) = w_0 + x \quad \text{at each cross-section } x. \tag{9}$$

The channel is shown in Fig. 7.

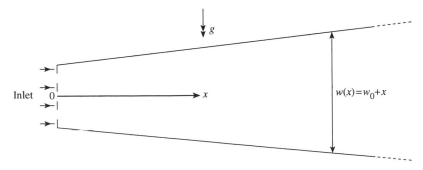

Fig. 7.   Expanding channel

To find $p(x)$ and $v(x)$ we return to equations (5) and (8), and now substitute for $w(x)$ by $w_0 + x$. We obtain

$$p(x) = p_0 + \tfrac{1}{2}\rho v_0^2 \left[1 - \frac{w_0^2}{(w_0 + x)^2} \right]$$

$$v(x) = v_0 \frac{w_0}{(w_0 + x)}$$

at each cross-section x down the channel. Observe that the pressure $p(x)$ increases whilst the speed $v(x)$ decreases as we move down the channel (increasing x). This should be expected from Theorem 1.

4. Blockage in a closed channel

We wish to examine the situation which arises when a central blockage occurs in a horizontal, closed channel of constant width w_0, as shown in Fig. 8.

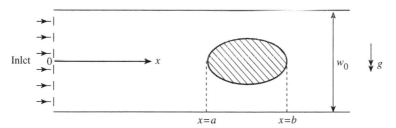

Fig. 8.   Obstructed channel

The blockage is *symmetric* about the centre line of the channel, and lies between $x = a$ and $x = b$. The width of the blockage is given by $d(x)$, when x lies between a and b ($a \leq x \leq b$). Because of the *symmetry* we need only consider half of the channel, as shown in Fig. 9.

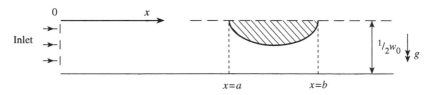

Fig. 9.   Half of obstructed channel

From equation (3) we know that at each cross-section in Fig. 9,

$$Q(x) = Q_0 = v_0 \times \tfrac{1}{2} w_0 \times b_0.$$

However, $$Q(x) = v(x) \times \tfrac{1}{2} (w_0 - d(x)) \times b_0$$

as we pass over the blockage. Therefore at cross-sections x between a and b,

$$v(x) \times \tfrac{1}{2} (w_0 - d(x)) \times b_0 = v_0 \times \tfrac{1}{2} w_0 \times b_0.$$

Cancelling gives $$v(x)(w_0 - d(x)) = v_0 w_0,$$

and so, $$v(x) = \frac{v_0 w_0}{(w_0 - d(x))}, \quad \text{for all } a \le x \le b. \tag{10}$$

The equation for the pressure is, from equation (7),

$$p(x) + \tfrac{1}{2} \rho v(x)^2 = p_0 + \tfrac{1}{2} \rho v_0^2 \quad \text{at each } x.$$

Substituting from equation (10) and re-arranging leads to

$$p(x) = p_0 + \tfrac{1}{2} \rho v_0^2 \left[1 - \frac{w_0^2}{(w_0 - d(x))^2} \right] \quad \text{for all } a \le x \le b. \tag{11}$$

We now have:

Theorem 2

In a horizontal channel of rectangular cross-section and constant width, the water speed increases as it passes around a blockage. Correspondingly, the water pressure decreases around a blockage.

Worksheet 1

1. Water is pumped into the inlet of a horizontal channel at pressure 10^4 Pascal and at speed $5\,\mathrm{ms}^{-1}$. The maximum width of the channel is $0.2\,\mathrm{m}$, the minimum width of the channel is $0.08\,\mathrm{m}$ whilst the width of the inlet is $0.1\,\mathrm{m}$.
   (a) Sketch how the channel may look.
      Now use the formula

$$p(x) = p_0 + \tfrac{1}{2} \rho v_0^2 \left(1 - \frac{w_0^2}{w(x)^2} \right), \tag{A}$$

$$v(x) = \frac{v_0 w_0}{w(x)}, \tag{B}$$

      to
   (b) find the highest water pressure in the channel.
   (c) find the lowest water pressure in the channel.
   (d) find the highest water speed in the channel.
   (e) find the lowest water speed in the channel.

2. Water flows through a horizontal channel whose shape is sketched below.

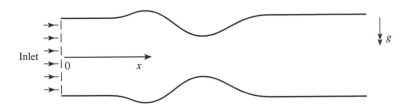

Mark on the figure, with a cross, those points on the channel surface where the speed is highest and, with a dot, those points on the channel surface where the pressure is highest. Explain why you have chosen these points.

3. Re-draw the above figure, and now mark, in the same way, the points where the speed and pressure are lowest. Again, explain your choices.

4. How will the highest and lowest values of speed and pressure from questions 2 and 3 compare with the inlet speed and pressure in the above figure?

5. Water is pumped into a horizontal, straight channel at pressure 10^4 Pascal and speed $5\,\mathrm{ms}^{-1}$ through an inlet of width 0.5 m. The channel is blocked by an obstacle, as shown below. The pressure at a distance x down the channel is then given by

$$p(x) = p_0 + \tfrac{1}{2}\rho v_0^2 \left[1 - \frac{w_0^2}{(w_0 - d(x))^2} \right]. \tag{C}$$

When the maximum width of the obstacle is 0.1 m, find the lowest pressure in the pipe. Mark, on the figure below, where this occurs on the surface of the pipe.

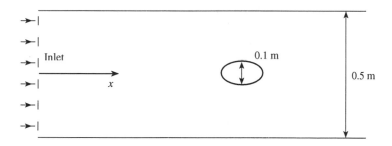

Outside the channel, the air is at pressure $10^5 \times 1.012$ Pascal. The material of the channel surface is only strong enough to sustain a pressure difference of $10^4 \times 9.5$ Pascal across the channel wall. Decide whether or not the channel wall will fracture.

5. Flow in non-horizontal closed pipes

Consider water flowing through a pipe of circular cross-section with variable radius. Let x be a coordinate measuring distance along the centre line of the pipe from the inlet at $x = 0$. The coordinate y measures distance in the vertical direction (upwards) with the inlet level at $y = 0$ as shown in Fig. 10.

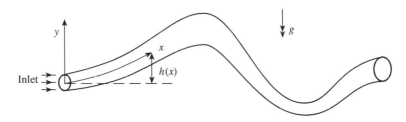

Fig. 10.   Circular pipe with varying radius and height

The *radius r* of the pipe varies as we move along, and so $r = r(x)$. Also the *height h* of the centre line of the pipe varies as we move along, and so $h = h(x)$.

Water enters the inlet of the pipe, which has radius r_0, with speed v_0 and at pressure p_0.

As before

$$Q_0 = Q(0) = \text{water speed at } (x = 0) \times \text{pipe cross-sectional area at } (x = 0)$$

$$= v_0 \times (\pi \times r_0^2) = v_0 \pi r_0^2. \tag{12}$$

At a distance x from the inlet

$$Q(x) = v(x)\pi r(x)^2. \tag{13}$$

However, because the water is incompressible,

$$Q(x) = Q_0 \text{ at each cross-section } x. \tag{14}$$

Substituting from equations (12) and (13) into (14) results in

$$v(x)\pi r(x)^2 = v_0 \pi r_0^2.$$

On cancelling π, this becomes

$$v(x)r(x)^2 = v_0 r_0^2 \text{ at each } x \tag{15}$$

which is the equation of conservation of mass for the flow of water in the pipe.

The second equation, obtained by application of Newton's second law of motion to the water in the pipe, is now modified. Since the centre line of the pipe is *no longer horizontal*, the force due to *gravity* now affects the flow in the pipe. The corresponding *Bernoulli equation* is that

$$p(x) + \tfrac{1}{2}\rho v(x)^2 + \rho g h(x) \tag{16}$$

remains the same (constant) at any cross-section x along the pipe. Therefore,

$$p(x) + \tfrac{1}{2}\rho v(x)^2 + \rho g h(x) = p_0 + \tfrac{1}{2}\rho v_0^2 \text{ at each } x. \tag{17}$$

Equations (15) and (17) are *two equations* to find the *two unknowns $p(x)$* and $v(x)$ at each cross-section x down the pipe from the inlet at $x = 0$. We have the following result.

Theorem 3

The water pressure $p(x)$ and the water speed $v(x)$ at each cross-section x in the pipe are given by

$$v(x) = \frac{v_0 r_0^2}{r(x)^2} \tag{18}$$

$$p(x) = p_0 - \rho g h(x) + \tfrac{1}{2}\rho v_0^2 \left(1 - \frac{r_0^4}{r(x)^4}\right). \tag{19}$$

Proof

To obtain equation (18) we simply divide both sides of equation (15) by $r(x)^2$. To obtain equation (19), we first subtract $\tfrac{1}{2}\rho v(x)^2 + \rho g h(x)$ from both sides of equation (17), and then substitute for $v(x)$ from equation (18). □

Example 3

A pipe has constant radius, so that $r(x) = r_0$ at all cross-sections of the pipe. Theorem 3 then shows that

$$v(x) = v_0$$
$$p(x) = p_0 - \rho g h(x)$$

at all cross-sections x along the pipe. Therefore, the water speed is constant along the pipe, whilst the water pressure depends only upon height above the inlet of the pipe. The *pressure decreases* with *increasing height*.

Example 4

We consider a tap supplied by water from a cylindrical tank shown in Fig. 11. The radius of the tap is r_T and the radius of the tank is r_0. The height of the tap above the bottom of the tank is H, whilst the depth of water in the tank is d. A piston is fitted into the tank and water is driven from the tank to the tap by depressing the piston at a constant speed v_0.

We should like to know how fast the piston must be depressed, and how much pressure we must apply to the piston to do this, so that the water emerges from the tap with a desired speed, say v_T.

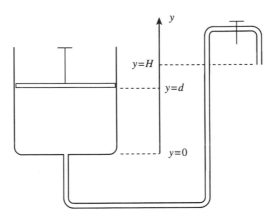

Fig. 11. Tap supplied by tank

We may regard the tank and pipe as a single pipe with varying radius. Therefore,

$$Q \text{ at the pump} = Q \text{ at the tap},$$

so that

$$v_0 \pi r_0^2 = v_T \pi r_T^2.$$

Cancelling π and dividing by r_0^2 gives

$$v_0 = \frac{v_T r_T^2}{r_0^2}, \tag{20}$$

as the speed with which the piston must be depressed.

We next apply the Bernouli equation (16) just below the piston and at the top outlet. This gives

$$p_0 + \tfrac{1}{2}\rho v_0^2 + \rho g d = p_A + \tfrac{1}{2}\rho v_T^2 + \rho g H. \tag{21}$$

Here p_0 is the water pressure just below the piston and p_A is the atmospheric pressure. Also

$$d = d_0 - v_0 t, \tag{22}$$

where d_0 is the height of water in the tank when the tap is turned on and t is the time over which the tap has been running. We substitute from equations (22) and (20) into (21), and rearrange, to obtain

$$p_0 = p_A + \tfrac{1}{2}\rho v_T^2 \left[1 - \frac{r_T^4}{r_0^4}\right] + \rho g H \left[1 - \frac{d_0}{H} + \frac{v_T}{H}\frac{r_T^2}{r_0^2} t\right]. \tag{23}$$

Now the piston moves down at a constant rate, so the pressure which needs to be applied on the top to achieve this must balance the pressure in the water just below. Therefore we must apply a pressure p_0, given in equation (23), to the piston. In many circumstances, the tap radius r_T is very much smaller than the tank radius r_0 (written $r_T \ll r_0$). In this situation $(r_T/r_0)^2$ is very much less than unity, and so, at least over a reasonable time interval, we may approximate equation (23) by

$$p_0 \approx p_A + \tfrac{1}{2}\rho v_T^2 + \rho g(H - d_0). \tag{24}$$

6. Applications

We consider some applications of the theory described in Sections 4 and 5.

6.1 Aircraft

Consider a wind tunnel with an aerofoil placed centrally. Air enters the tunnel at atmospheric pressure and flows around the aerofoil, as shown in Fig. 12.

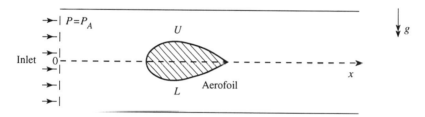

Fig. 12. Symmetric aerofoil

Although the fluid is no longer water, the theory of Section 4 still applies. When the aerofoil shape is symmetric about the centre line of the wind tunnel, the formula (11) shows us that the pressure on the upper surface (U) and the lower surface (L) of the aerofoil must be the same. We have no *lift* on the aerofoil.

To acquire lift we must design the aerofoil so that the pressure at U is lower than the pressure at L. How can we do this? We must make $d(x)$ in formula (11) *larger* at U and *smaller* at L. A design might be as shown in Fig. 13.

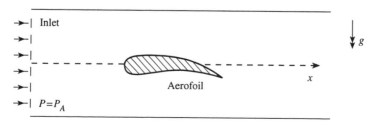

Fig. 13. Cambered aerofoil

This is by no means the whole story, but gives some indication as to why aerofoils need to be shaped (cambered) as in Fig. 13.

6.2　House roofs

It is well known that in high speed winds house roofs may be torn away from the structure. Why does this happen? Consider the wind tunnel experiment shown in Fig. 14.

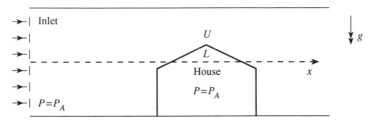

Fig. 14.　Roof lift-off

As the wind passes over the house roof, formula (11) determines that the pressure on the *outside* of the roof must fall below P_A, whilst the pressure in the house is at P_A. So the pressure at U is lower than the pressure at L, and this generates a *lift* on the roof.

6.3　Racing cars

Racing cars use air flow to acquire 'suction' which helps in holding them to the track. 'Skirts' are attached to the lower part of the body work to reduce the distance between the chassis and the track, as shown in Fig. 15.

To understand this effect, we imagine the car is stationary in a wind tunnel, with air flowing in through the inlet and around the car. Using formula (11), the pressure at U must be higher than the pressure at L, and so the car is 'pressed' onto the track.

Worksheet 2

A circular pipe has constant radius of 0.01 m, and has shape as shown below.

Water enters the pipe at the inlet with speed $2\,\mathrm{ms}^{-1}$ and at pressure 10^4 Pascal. When x measures distance along the centre of the pipe from the inlet, the water speed and pressure at a distance x along the pipe are given by

$$v(x) = \frac{v_0 r_0^2}{r(x)^2} \tag{A}$$

$$p(x) = p_0 - \rho g h(x) + \tfrac{1}{2}\rho v_0^2 \left(1 - \frac{r_0^4}{r^4(x)}\right) \tag{B}$$

Fig. 15. Down force on a racing car

where v_0, p_0, r_0 are the water speed, water pressure and pipe radius at the inlet, and $g \approx 10\,\text{ms}^{-2}$ is the acceleration due to gravity.

1. Mark on the Figure below where the highest and lowest water pressure occur. Explain your choice.

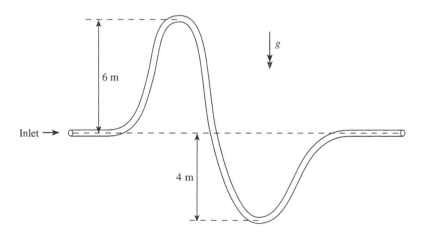

2. Calculate the highest and lowest water pressure in the pipe.
3. How does the water speed change as it flows along the pipe?
4. Draw on the axes below how you think the water pressure will change as it flows along the pipe.

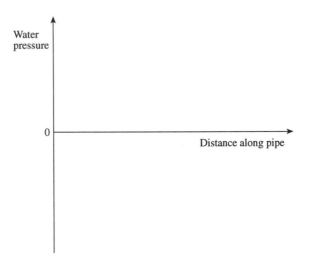

7. Open channel flow

Here we consider water flowing over a fixed bed, with the upper surface open to the atmosphere. Examples are flows in rivers and canals. Far upstream the river bed is flat and horizontal and the water is flowing with speed v_0 and depth h_0, as shown in Fig. 16.

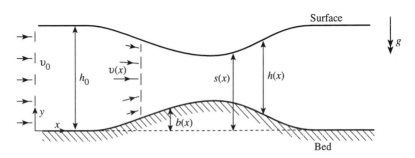

Fig. 16. Flow over a hump

The river encounters a hump or depression in the bed on moving downstream. How does the surface respond? We will investigate this. Let x measure distance (from upstream) horizontally, and y measure distance vertically, as shown in Fig. 16. We introduce the following symbols:

$$y = s(x) \quad \text{gives water surface level}$$
$$y = b(x) \quad \text{gives bed level}$$
$$h(x) \quad \text{gives water depth}$$
$$v(x) \quad \text{gives water speed}$$

at each distance x downstream. We assume the flow in the river is *steady*, so that conditions at any x do not change in time.

Since the water in the river is incompressible and the flow is steady, the volume of water entering upstream over one second must be the same as the volume of water leaving through the cross-section at x over one second. When w_0 is the constant width of the river, this requires

$$v(x) \times h(x) \times w_0 = v_0 \times h_0 \times w_0 \quad \text{at all cross-sections } x.$$

Cancelling w_0 gives

$$v(x)h(x) = v_0 h_0 \quad \text{at all cross-sections } x. \tag{25}$$

This represents conservation of water mass in the river. Observe that higher water speeds lead to shallower water depths.

For river flows the corresponding Bernoulli equation takes the form

$$\tfrac{1}{2}v(x)^2 + g(h(x) + b(x)) = \tfrac{1}{2}v_0^2 + gh_0 \quad \text{at all cross-sections } x. \tag{26}$$

The equations (25) and (26) need to be solved (once we know the bed shape $b(x)$) to find the river depth $h(x)$ and the river speed $v(x)$ at each cross-section x downstream. This leads to a complicated cubic equation for $h(x)$. However, when the hump or depression is small compared to the overall depth, equations (25) and (26) may be simplified. They become

$$v_0(h(x) - h_0) + h_0(v(x) - v_0) = 0 \tag{27}$$
$$v_0(v(x) - v_0) + g(h(x) - h_0 + b(x)) = 0 \tag{28}$$

at each cross-section x. We re-arrange equation (27) to give

$$v(x) = v_0 - \frac{v_0}{h_0}(h(x) - h_0).$$

Now substitute for $v(x)$ into equation (28), which leaves

$$-\frac{v_0^2}{h_0}(h(x) - h_0) + g(h(x) - h_0 + b(x)) = 0.$$

We can now solve this equation to find $h(x)$ as

$$h(x) = h_0 + \frac{b(x)}{\left(\dfrac{v_0^2}{gh_0} - 1\right)} \tag{29}$$

at each downstream cross-section x. We have from equation (29) the following.

Theorem 4

(i) When the river flow encounters a small (relative to depth) hump in the bed, the depth *decreases* when $v_0^2 < gh_0$ and *increases* when $v_0^2 > gh_0$.
(ii) When the river flow encounters a small (relative to depth) depression in the bed, the depth *increases* when $v_0^2 < gh_0$ and *decreases* when $v_0^2 > gh_0$.

When $v_0^2 = gh_0$, the simplified theory fails and we must return to the original equations (25) and (26). A river which has $v_0^2 < gh_0$ is called *subcritical* whilst a river which has $v_0^2 > gh_0$ is called *supercritical*. The flow of both a supercritical and subcritical river over a small hump in the bed is illustrated in Fig. 17.

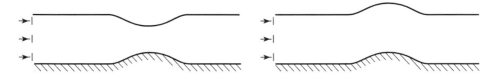

Fig. 17.   Subcritical and supercritical flows over a small hump

Solutions to Worksheet 1

1. (a) The channel may have the form shown below:

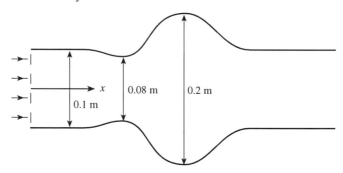

(b) From equation (A), highest water pressure occurs when the width $w(x)$ is largest. This is when $w = 0.2\,\text{m}$. This gives, after substitution into equation (A) (with $w_0 = 0.1\,\text{m}$, $v_0 = 5\,\text{ms}^{-1}$, $P_0 = 10^4$ Pascal), the highest water pressure as

$$10^4 + \tfrac{1}{2} \times 10^3 \times 5^2 \times \left(1 - \frac{(0.1)^2}{(0.2)^2}\right)$$

$$= 19\,375\,\text{Pascal} = 1.9375 \times 10^4\,\text{Pascal}$$

(c) The lowest water pressure occurs when $w(x)$ is smallest. This gives the lowest water pressure as

$$10^4 + \tfrac{1}{2} \times 10^3 \times 5^2 \times \left(1 - \frac{(0.1)^2}{(0.08)^2}\right)$$

$$= 10^4 - 7.031\,25 \times 10^3$$
$$= 2968.75 \text{ Pascal}$$
$$= 2.968\,75 \times 10^3 \text{ Pascal}.$$

(d) The highest water speed occurs, from equation (B), when $w(x)$ is smallest. Putting $w = 0.08$ m into equation (B) gives the highest water speed as

$$\frac{5 \times 0.1}{0.08} = 6.25\,\text{ms}^{-1}.$$

(e) The lowest water speed occurs when $w(x)$ is largest. The lowest water speed is, from equation (B),

$$\frac{5 \times 0.1}{0.2} = 2.5\,\text{ms}^{-1}.$$

2.

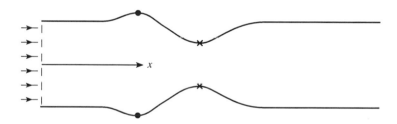

From equation (B), speed is highest when width is smallest. From equation (A), pressure is highest when width is largest.

3.

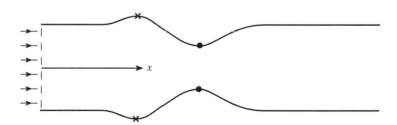

From equation (B), speed is lowest when width is largest. From equation (A), pressure is lowest when width is smallest.

4. The highest pressure is higher than the inlet pressure. The lowest pressure is lower than the inlet pressure. Similarly for the speeds. This follows from equations (A) and (B) since the largest width is larger than the inlet width and the smallest width is smaller than the inlet width.
5. In equation (C),

$$p_0 = 10^4 \text{ Pascal}, \ v_0 = 5 \text{ ms}^{-1}, \ w_0 = 0.5 \text{ m}.$$

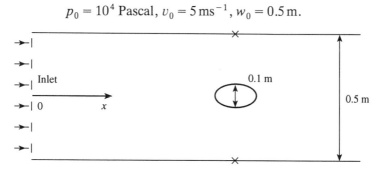

The lowest water pressure occurs when the width of the obstacle is largest. This is when $d(x)$ is 0.1. Therefore the lowest pressure is, using equation (C),

$$10^4 + \tfrac{1}{2} \times 10^3 \times (5)^2 \times \left[1 - \frac{(0.5)^2}{(0.4)^2} \right]$$

$$= 2968.75$$
$$= 2.968\,75 \times 10^3 \text{ Pascal}.$$

This occurs at points (X) on figure above. The pressure difference at this point is

$$10^5 \times 1.012 - 10^3 \times 2.968\,75$$
$$= 9.823\,125 \times 10^4 \text{ Pascal}.$$

The channel wall will therefore be expected to fracture.

Solutions to Worksheet 2

1.

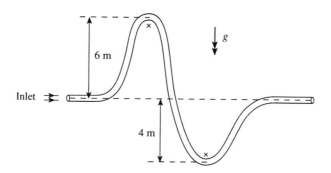

This follows from equation (B) which becomes, when radius is constant,

$$p(x) = p_0 - \rho g h(x). \tag{C}$$

2. Highest pressure occurs when pipe is at lowest level, and $h(x) = -4\,\mathrm{m}$. Substitute into equation (C) to get highest pressure as

$$10^4 + 10^3 \times 10 \times 4$$
$$= 10^4 \times (1 + 4)$$
$$= 5 \times 10^4 \text{ Pascal.}$$

Lowest pressure occurs when pipe is at highest level, and $h(x) = 6\,\mathrm{m}$. Substitute into equation (C) to get lowest pressure as

$$10^4 - 10^3 \times 10 \times 6$$
$$= 10^4 \times (1 - 6)$$
$$= -5 \times 10^4 \text{ Pascal.}$$

3. Water speed $v(x)$ is given by equation (A). Here $r(x)$ is equal to r_0 at all distances x down the pipe, so,

$$v(x) = v_0 = 2\,\mathrm{ms}^{-1}$$

at all distances x down the pipe.

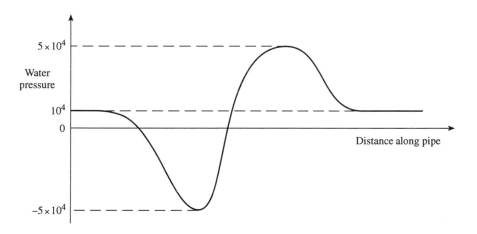

This follows from the diagram of the pipe and equation (C).

Acknowledgement

The photograph in Fig. 15 was kindly provided by Williams Grand Prix Engineering Ltd. of Wantage, UK.

8 Balloons and Bubbles

by MICHAEL SEWELL

Department of Mathematics, University of Reading

1. Experiments and questions

(a) Try to blow up a rubber party balloon, using only your lungs instead of a balloon pump. When the balloon is of the kind which becomes pseudo-spherical or pear-shaped, what significant experimental fact do you notice during the inflation process?

You will usually find that inflation is difficult *at first*, and becomes easier in the later stages. We shall look for a mathematical expression of this fact. You might have expected the opposite effect, that it would be progressively harder to inflate the balloon because the rubber surface has to be stretched more and more. The reality is more subtle, because opposing factors have to be balanced, as we shall see.

(b) If a balloon could be truly spherical after the slack is taken up, will it remain so during inflation? Very surprisingly, the answer is: *not necessarily*. You might expect nature to take the view that 'the sphere is the perfect shape' when there are no imperfections to cause asymmetry to develop. But nature has a richer agenda, and does not take that view. We shall demonstrate departure from sphericity with a meteorological balloon, of a kind used for sending instruments aloft. Such balloons can be truly spherical at first, apart from the nozzle. Again, a mathematical graph can describe the behaviour.

(c) What happens when you blow up a cylindrical balloon? The inflation begins as a pseudo-spherical bulge at one end, which then *extends* down the length of the balloon without much increase in width. Can we explain that mathematically? You might have expected the whole cylinder to grow gradually from a small cylinder to a large one, but it does not behave like that.

(d) We shall blow two soap bubbles, of nearly the same spherical size, at opposite ends of a glass tube. What would you expect to happen when a valve is opened to allow air to pass between the bubbles? Which way will the air flow? Will the bubbles finally have the same size, or different sizes? We shall give some mathematical theory.

2. Equilibrium

The search for explanations of such effects must begin with a discussion of *forces*, and of how different forces can balance each other to create what is called *equilibrium* of whatever they act upon, so that it is not moving.

Equilibrium of any patch of a balloon or bubble is a balance between

(a) excess air pressure = internal pressure minus atmospheric pressure, and
(b) membrane tension in the skin, which is ultimately due to forces between the
   molecules of rubber or soap film.

The weight of the balloon or bubble, which is the force exerted by gravity, is
much smaller than (a) or (b), and therefore we neglect it.

Figure 1 shows a cross-section through a patch of balloon. The curve is
symmetrical about the vertical on the page, but otherwise of any smooth shape.
Forces are represented, in magnitude and direction, by the length and direction
(respectively) of arrows. Figure 1 includes (a) the *resultant*, or combined effect,
P of the excess air pressure, which actually acts on the whole surface of the
patch as shown schematically in the two equivalent sets of forces in Fig. 2.

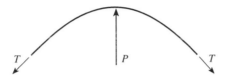

Fig. 1.   Forces on a patch of balloon

Fig. 2.   Pressures and their resultant *P*

Figure 1 also shows (b) the tension forces *T* at each edge of the patch. They act
tangentially to the surface, and are equal by symmetry. Each represents the
intermolecular forces exerted *by* the rubber outside the patch, *on* the rubber
patch itself, across the edge.

Anything which requires both a magnitude and a direction to specify it fully is
called a *vector*. Force is an example of a vector (and velocity is another
example). This is why newspapers speak of the vectored thrust provided by each
rotatable engine of the Harrier jump jet.

When a vector is represented as an arrow, we can always draw any right-
angled triangle using that arrow as the hypotenuse. In the geometry of vectors it
is a hypothesis, called the *vector law of addition*, that that arrow is equivalent to
the pair of arrows along the other two sides, as shown in Fig. 3. *V* and *H* are
called the vertical and horizontal *components* of *T*.

Fig. 3.   Equivalence of vectors

Applying this decomposition twice to the T forces in Fig. 1 shows that the (upward) pressure forces balance the two (downward) vertical tension components if

$$P = 2V. \tag{1}$$

This is the *equilibrium equation* for the patch. The two horizontal tension components are equal and opposite, and so balance by symmetry.

3. Pressure on a patch

The *pressure* on a surface is defined as the *force per unit area* acting perpendicular to that surface. Atmosphere pressure on the surface of the sea or earth is an example. Look at the scale on a barometer. Low pressure on a mercury barometer would support a column of 28 inches of mercury, while high pressure supports 31 inches. In the modern units used in the weather forecast, atmosphere pressure near sea level is normally within the range 975 ± 50 millibars.

Pressure on a *curve*, like that on the left in Fig. 2, is defined as force per unit *length* acting perpendicular to that curve.

Suppose that a straight chord of length $2c$ joins the ends of the curved cross-section of the patch in Fig. 1, as shown in Fig. 4. We can think of the curve as being made up of a sequence of small distances s such as are approximated by the hyptenuses of small right-angled triangles like that shown, having horizontal and vertical sides x and y respectively. Let p denote the local excess pressure, providing a force ps perpendicular to s as shown in Fig. 4. Because there is no motion, it is reasonable to assume that p is *uniform*, i.e. it has the same value at every place on the curve.

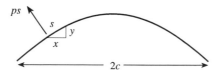

Fig. 4.   Curve under local excess pressure p per unit length

By the vector law of addition, this force is equivalent to the combined horizontal and vertical forces py and px shown in Fig. 5(a), such that the

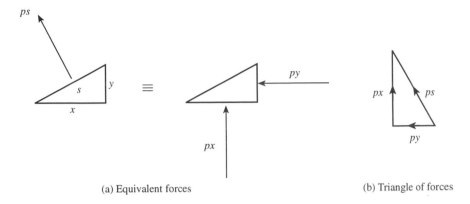

(a) Equivalent forces   (b) Triangle of forces

Fig. 5.   Components of pressure forces

so-called *triangle of forces* in Fig. 5(b) closes. The triangle in Fig. 5(b) has the
same angles as that in Fig. 5(a), but it is rotated through a right angle and scaled
in size by the factor p.

We now add the forces on each such small triangle on the right of Fig. 5(a),
along the whole curve. The horizontal forces cancel out in pairs by symmetry, as
Fig. 6 shows.

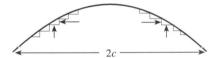

Fig. 6.   Addition of pressure forces along the curve

However, the sum of the vertical forces is all upward and of amount
$$\Sigma px = p\Sigma x = p2c. \tag{2}$$
Here Σ denotes 'the sum of all such things as', for example $\Sigma x = 2c$ even
though individual x-values may change from one small triangle to the next; but
p is the same for each triangle, and therefore a common factor of the terms on
the left. (This process of adding small quantities is called *integration* in mathematics. Rules have been developed to shorten it, but do not need them here.)

Theorem 1

The pressure required to maintain the curve in Fig. 1 in equilibrium is
$$p = \frac{V}{c} \tag{3}$$
in terms of the vertical tension component V and the half-chord width c.

Proof
From the definition of the resultant in Fig. 2, $P = 2pc$ in equation (2). Using
equation (1) then gives equation (3). $\square$

The open box signifies the end of a proof. Its use is a modern convention. In older books QED is used, standing for 'Quad erat demonstrandum' in Latin, which means 'Which was to be proved.'

In the particular case when the cross-section through the patch of the balloon considered is an arc of a circle of radius R, as can be the case when the balloon is a circular cylinder, we can express p in terms of T and R instead of V and c as follows.

Theorem 2

The pressure required to maintain a circular arc of radius R in equilibrium is related to the internal surface tension in the membrane by

$$p = \frac{T}{R} \qquad (4)$$

regardless of the span of the arc.

Proof

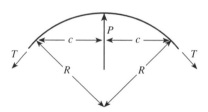

Fig. 7. Forces on a circular arc

The resultant forces and dimensions are shown in Fig. 7. Because the tension T is tangential to the circle, and therefore perpendicular to the radius, the right-angled triangle with hypotenuse R has the same shape as the triangle of forces in Fig. 3, even though it is scaled in size and rotated through a right angle. Therefore the ratios

$$\frac{c}{R} = \frac{V}{T} \qquad (5)$$

are the same. Hence $V/c = T/R$, which proves equation (4) from equation (3).
□

Triangles which are the same shape, but of different size, are called *similar*. The solutions to Worksheet 1 include an alternative and shorter proof of equation (4). (The ratio in equation (5) will be recognized as the sine of the angle opposite c or V, by those acquainted with trigonometry, but I have been able to avoid that subject in favour of other basic ideas.)

What are the pros and cons of equations (3) and (4)? For (3) the shape of the curve need not be circular, as it must for (4); but in (4) the whole of the tension T appears, and not just one component of it, as in (3). The tension T acting within a circular arc is sometimes called the *hoop stress* and, like R, it has a basic importance which is independent of the *extent* of the arc.

4. Equilibrium of a spherical patch

Thus far we have considered the equilibrium of a curve in the plane of the paper, but that is not realizable experimentally except in special cases, such as the cross-section of a cylindrical balloon. Balloons and bubbles are curved surfaces in three-dimensional space, so we need to ask, for example, what the equilibrium formula (4) becomes when the number of dimensions is increased from two to three. Which of the following guesses for a spherical balloon would you choose:

$$p = \left(\frac{T}{R}\right)^2, \quad p = \frac{T}{2R}, \quad p = \frac{2T}{R}, \quad \text{or} \quad p = \frac{T}{R} \quad \text{again?}$$

Here p is now force per unit area (instead of per unit length) as explained at the beginning of Section 3, and T becomes the force, tangential to the surface, per unit length of a curve lying in the surface, and perpendicular to the curve. Informed guessing of what we should try to prove is a useful tactic in the creation of new mathematics.

Theorem 3

The pressure required to maintain a spherical patch of radius R in equilibrium is

$$p = \frac{2T}{R}. \tag{6}$$

This is the correct result, and a proof is given among the solutions to Worksheet 1. The result is locally valid even if the *whole* surface, of which the patch is a part, is not a sphere. Frequently only part of a surface is spherical, for example because of the presence of a nozzle or other imperfections.

The results obtained so far apply just as well to a soap bubble as to a rubber balloon. The difference between them will be seen when we consider how T depends on R. We have not needed this dependence yet.

Worksheet 1

1. If three equal forces in the same plane pull on a point, what is the angle between them if the point is kept in equilibrium?
2. A circle of radius R is imagined to be filled with fluid which exerts a pressure p per unit length perpendicular to any curve within it. A surface tension T tangential to the circle maintains equilibrium with the pressure. By considering the forces on the semi-circle shown in the diagram, prove that the equilibrium equation is $p = T/R$.

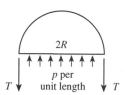

3. The boundary of a closed plane region consists of a curve whose ends are joined by a straight line of length $2c$, such that the region is symmetrical about the perpendicular bisector of the line. The region is filled with fluid which holds the boundary in equilibrium by exerting a uniform pressure p per unit length perpendicular to every part of the boundary, as shown.

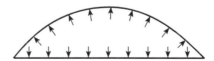

By considering the forces on the chord, construct an argument which proves that the forces on the curve will have a vertical resultant $2cp$.

4. A sphere of radius R is filled with fluid which exerts a pressure p per unit area perpendicular to any surface within it. A membrane tension T per unit length acts perpendicularly across any curve within the spherical skin, and tangential to the skin, and maintains equilibrium with the pressure. By considering the forces on the hemisphere shown in the diagram, prove that the equilibrium equation is $p = 2T/R$.

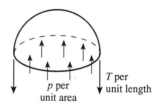

p per unit area

T per unit length

5. A sphere of radius R is cut in two by a plane, which intersects the sphere in a circle of radius $c < R$. The smaller of the two parts is called a spherical cap. The cap is now regarded as a vessel filled with fluid which exerts a pressure p per unit area perpendicular to every part of its boundary. By considering the forces on the circular plane prove that, in equilibrium, the forces on the curved part of the cap will have a resultant $\pi c^2 p$ along the axis of symmetry.

6. What is the length of the circumference of the cap in Question 5? Let T per unit length be the surface tension within the spherical surface, acting tangentially to the surface on any curve lying in it, as explained just before Theorem 3. Prove that the components of the total surface tension acting on the circular boundary of the cap are zero and $2\pi c^2 T/R$ perpendicular and parallel to the axis of symmetry, respectively.

7. By combining the results of Questions 5 and 6, prove that equilibrium of the cap requires $p = 2T/R$, thus proving Theorem 3 in a different way to that in the text or in Question 4.

8. The formula $p = 2T/R$ describes a surface in a three-dimensional space for positive values of cartesian coordinates p, T and R. This could be called an

equilibrium surface for all possible spherical balloons, i.e. the geometrical description of all possible equilibrium combinations of p, T and R. Construct a picture of the surface by any means available to you, such as free-hand sketches, a calculator or computer. Cross-sections having different fixed R, or different fixed T, can be helpful.

5   Tension in a soap film

A soap film really consists of a thin layer of soap solution having an interface with the air on either side. Different types of soap solution are available, and their chemistry is quite complicated. Each of the two surfaces will have a surface tension, and they combine together to provide an average film tension. For our mathematical purpose it is adequate to regard this as the membrane tension T in an idealized single surface of negligible thickness which represents the soap film.

Physics shows that T can depend on position, and on velocity if there is significant movement within the film, and that it can depend strongly on temperature. Here we shall ignore these variations. This is a common mathematical approximation, or idealization, for studying the equilibrium of soap bubbles.

In particular we assume that, at any point of a soap film, T is the same in all directions tangential to the film. Because the soap film is made of fluid, this may be regarded as an analogy of the assumption that the pressure at a point within a three-dimensional fluid is the same in all directions.

Another reasonable assumption about T is that, if the concentration of the soap solution is sufficiently large, then for a spherical bubble in equilibrium T *is independent of R*. Writing this assumption into equation (6) in the form $pR = 2T$ implies that

$$pR = \text{constant}. \tag{7}$$

The assumption begins to fail for low concentrations.

6. Equilibrium of a patch of general shape

It is proved in geometry books that, at each point of a smooth surface of any shape, the following property holds. Imagine the plane which is tangential to the surface at the considered point, i.e. the tangent plane. Every plane through that point which is perpendicular to that tangent plane will intersect the surface in a curve. Among the curvatures of all these cross-sections, one will be a *maximum* and another will be a *minimum*, with associated radii of curvature R_1 and R_2 (i.e. *as if* the cross-sections are locally approximated by *circles* with these radii); and the planes which make these two special cross-sections are at *right angles* to each other. We say that R_1 and R_2 are the *principal* radii of curvatures at that point of the surface. They will have opposite signs if the curvatures are in opposite directions. Another point on the surface will have another such pair of principal radii, but often with a different pair of values.

Theorem 4

When a local patch of soap film or balloon is in equilibrium, the excess pressure p across it and the membrane tension T within it are related to the principal radii of curvature by

$$p = T\left(\frac{1}{R_1} + \frac{1}{R_2}\right).$$ (8)

It is usually said that this formula was proved by T. Young in 1805 and P. S. Laplace in 1806. We shall not prove it here, but you can illustrate it by the home experiment described in Section 7.

The hypothesis of constant T mentioned in Section 5 can be shown to imply that the surface *area* of a soap film which will connect to any given boundary is minimized when the film is in equilibrium, among all possible shapes of soap film which might be in motion when joined to the boundary. This minimum area property can be proved to imply equation (8), and it has been extensively explored by mathematicians.

Notice two special cases of equation (8). For a sphere of radius R, every cross-section perpendicular to a tangent plane has the same radius, so $R_1 = R_2 = R$, and equation (8) becomes equation (6). For a circular cylinder of radius R, the minimum principal radius is $R_1 = R$ for a cross-section transverse to the cylinder, and the maximum principal radius is $R_2 = \infty$ for a cross-section parallel to the axis of the cylinder, so that equation (8) becomes equation (4).

7 Equilibrium under zero excess pressure

Soap films can provide a convenient illustration of surfaces which are in equilibrium under zero excess pressure. This occurs when the pressure is atmospheric on *both* sides of the film. The left side of (8) is then zero, and therefore, because $T \neq 0$,

$$\frac{1}{R_1} + \frac{1}{R_2} = 0.$$ (9)

Theorem 5

There are three locally possible equilibrium shapes for a surface across which the excess pressure is zero, as follows.

(a) Flat, with R_1 and R_2 both infinite.
(b) Saddle-shaped, with R_1 and R_2 both finite, and equal in magnitude, but opposite in sign.
(c) Junction of two flat parts, or of two saddle shapes, or of one of each.

Proof
It is clear that (9) is satisfied either where (a) applies, or where (b) applies
($R_1 = -R_2$), or where (c) applies, and not otherwise.   □

A patch of surface in the neighbourhood of a point where R_1 and R_2 have
opposite signs is called *saddle-shaped*, and (b) is a particular case of this.

You can perform the following experimental illustrations of Theorem 5 in the
kitchen. Plastic-covered garden wire, of about 1 mm diameter, can be bent to
any desired shape, and used to provide a soap film surface by dipping it into
dishwashing detergent, such as neat Fairy Liquid.

(a) Form the wire into any closed flat loop, such as an approximate circle. The
    film which emerges is also flat, i.e. a plane. The result is not surprising, but
    Theorem 5 does predict it.
(b) Form the wire into any single closed loop which is *not* flat, such as the skew
    quadrilateral formed by the four approximately straight lines in Fig. 8. The
    film which emerges is saddle-shaped *at each point* (not just at the central
    point where one might imagine sitting on the saddle).

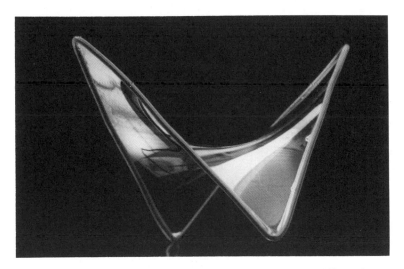

Fig. 8. Saddle-shaped surface bounded by skew quadrilateral

(c) Form the wire into two parallel circles near to each other, but capable of
    being moved apart. The film which emerges first has a saddle-shaped part
    connecting the two circles, *together with* a flat part joined to the waist of the
    saddle-shaped part as shown in Fig. 9. It is best to begin with the wire circles
    very close to each other, and then to move them slowly apart after the film
    has been established.
(d) If the central flat part is burst with a finger or a pencil, there remains the
    saddle shaped film alone connecting the two circles, as shown in Fig. 10.

Fig. 9. Saddle-shaped surface joined by flat surface

Fig. 10.   Catenoid bounded by two loops

This particular shape of surface is called a *catenoid*, because its cross-section on a plane through the centres of the wire circles (like the plane of the paper) is a *catenary*; the latter is the shape adopted by a uniform string or wire suspended under gravity from two points (e.g. a washing line or telephone wire).

(e) As the two circles are moved gradually further apart, the waist of the catenoid becomes narrower and it is increasingly difficult to hold it in stable equilibrium. Eventually the film jumps suddenly to leave one flat film across each of the two circles. This happens whether or not the central part had previously been burst artificially as in (d).

8. Connected soap bubbles

Here we describe the experiment mentioned in Section 1(d). The glass apparatus is shown in Fig. 11, with a bubble of similar size (about 3 cm diameter) at each end. Neat washing-up detergent is too viscous for this purpose. It is better to use, for example, a bubble solution of Teepol:glycerol:water in the ratio 1:8:32 by volume. The bubbles last longer if the solution has recently been in a refrigerator.

The essentials of the apparatus are shown in Fig. 12. A branched tube has three open ends A, B, C. Each end has a tap which can be opened or closed. The tap at C can be opened to a pressurized reservoir (for example, via a rubber tube to your lungs). Open A, close B, blow a bubble at A, and close A. Next open B, blow a bubble of similar size there, and close B. Now close C. Each bubble is blown after dipping the end A or B into a raised dish of bubble

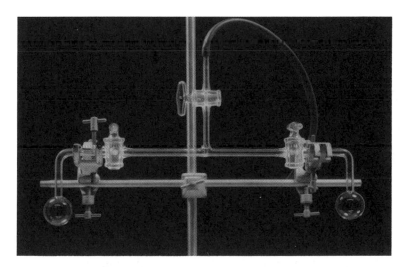

Fig. 11. Apparatus for connecting soap bubbles

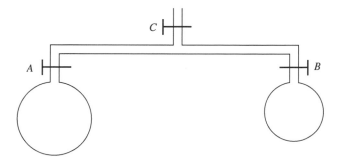

Fig. 12. Schematic version of Fig. 11

mixture to create a film over the end. It is best if the ends of the tube are clean. Both bubbles will need to co-exist for a minute or two at least. Air currents in the vicinity do not help, and neither does a warm atmosphere.

When both bubbles have been created, tap C is kept closed for the remainder of the experiment, and taps A and B are initially closed. Suppose that each bubble is part of a sphere terminated by the circular end of the tube, and that the membrane tension T is the same in each. Also suppose, as indicated in Section 5, that in any subsequent changes this value of T remains constant and that each bubble continues to be part of a sphere, although its radius may change.

What equilibrium sizes are possible according to the equilibrium equation $p = 2T/R$ in (6)? By equation (7) we must always have that

$$p_A R_A = p_B R_B \qquad (10)$$

for the two bubbles at A and B. It is most unlikely that the bubbles will have been created with exactly the same radius. Suppose, therefore, that $R_A > R_B$ as in Fig. 12. Then $p_A < p_B$ by (10). This means that, if the taps at A and B are opened, the air will flow from B to A until the pressures are equalized. This answers one of the questions in Section 1(d). Applying equation (10) in the new equilibrium situation with $p_A = p_B$ implies $R_A = R_B$.

Theorem 6

There are *two* possible final equilibrium configurations for the pair of connected bubbles, as follows and shown in Fig. 13.

(a) Each bubble is the same size.
(b) One bubble is much larger than the other, in such a way that the smaller one is a spherical cap (smaller than a hemisphere) of exactly the size which could fit onto the larger one (larger than a hemisphere) around a circle which has the same circumference as the tube.

Proof
When the bubbles are connected, the air will flow until the pressures are equalized in equilibrium, as explained above, and then the radii must also be the same. This is achieved not only in case (a), but also in case (b) which we can call *little and large*.   □

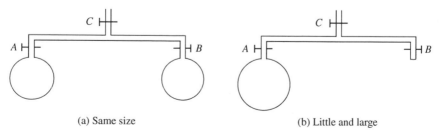

(a) Same size   (b) Little and large

Fig. 13.   Possible equilibria of connected bubbles

Which of these two solutions do you expect to see in the experiment, when the taps A and B are both opened after blowing the bubbles? It turns out that the system always goes to the 'little and large' solution (b). Because $R_A > R_B$ implies $p_B > p_A$, this initial pressure difference causes the bigger radius to grow and the smaller one to decrease. This answers the other question in Section 1(d). The change happens quite slowly, but it will happen repeatedly.

It is the case, although we shall not prove it mathematically here, that the 'same size' solution (a) is *unstable*, and it will never be seen experimentally after the taps are opened. If it did exist fortuitously when the bubbles are first blown, the smallest perturbation would cause runaway to the 'little and large' solution (b) after the taps are opened.

9. Tension in rubber

A rubber sheet has a very different molecular structure from that of a soap film. This means that it will extend in a very different way under membrane tension. Rubber is an elastic material composed of a cross-linked network of flexible long-chain molecules. They act in combination to provide some resistance to an external tension.

The simplest deformation of rubber occurs in the straight stretching of an elastic band. Under zero tension it has a natural straight length. Under nonzero tension we define the *stretch* to be

$$\frac{\text{current length}}{\text{natural length}} \geq 1.$$

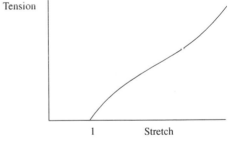

Fig. 14. Straight elastic response

Tension is force per unit area of cross-section in this case, and experiment shows that its dependence on stretch is an S-shaped graph as in Fig. 14.

A curved sheet such as a rubber balloon can be regarded as an aggregate of many elastic bands. Because of the curvature, however, their mutual interactions are not a simple superposition of response curves like Fig. 14, which they would be for two or more bands connected in a straight line. Instead there is a weakening effect which manifests itself in the graph relating membrane tension T to radius R, as follows.

A spherical rubber balloon has a slack or natural radius, say r, before the tension is applied. When the stretched radius is R, the stretch in a great circle is $2\pi R/2\pi r = R/r \geq 1$. The dependence of T upon R during inflation is shown qualitatively by the graphs in Fig. 15, for two different types of rubber (recall from Section 7 that T is independent of R for a soap bubble, so that the graph would be a straight horizontal line over the significant range).

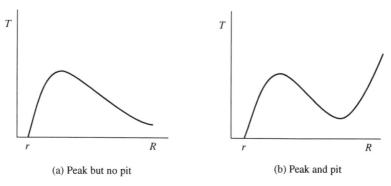

(a) Peak but no pit   (b) Peak and pit

Fig. 15. Dependence of T upon R for different rubber balloons

10. Pressure peaks and pits

The dependence of T upon R which is represented by a graph like those in Fig. 15 is called a *function*, and written as $T(R)$. Dividing it by $\frac{1}{2}R$ implies another function $p(R)$, which describes the dependence of p upon R in equilibrium according to the equation $p = T(R)/\frac{1}{2}R = 2T/R$ in (6) for the equilibrium of a sphere. Here we deduce the main features of the graph of $p(R)$ from Fig. 15 and from $p = 2T/R$.

Theorem 7

If we draw a straight line from the origin to touch the $T(R)$ graph in Fig. 15, there will be one such tangency point in Fig. 15(a), but one on each of two lines in Fig. 15(b).

(a) The tangency point which precedes the local maximum of $T(R)$ will specify a *local maximum* of $p(R)$, i.e. a pressure *peak*.
   The pressure peak occurs before the tension peak in Fig. 15(a) and (b).
(b) The tangency point which occurs after the local minimum of $T(R)$ will specify a *local minimum* of $p(R)$, i.e. a pressure *pit*.
   The pressure pit occurs after the tension pit in Fig. 15(b).

Proof

(a) Choose a constant k, and draw a straight line from the origin having slope k and therefore local height kR. If k is low enough this will intersect Fig. 15(a) in two places, as the chord shown in Fig. 16 (top left). At these two R values, $T = kR$ and therefore $p = 2k$, using $p = 2T/R$, so that the two pressures will have the same value as shown in Fig. 16 (bottom left). Now imagine k to increase gradually until the two intersection points of the chord with the curve come together. This will define a single tangency point on $T(R)$ as shown in Fig. 16 (top right), at a value of R corresponding to a local maximum or pressure peak of $p(R)$ as shown in Fig. 16 (bottom right).

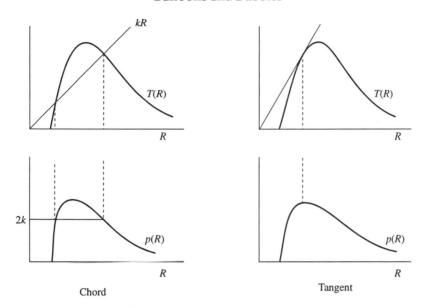

Fig. 16. Pressure peak construction

(b) Choose another value of k and repeat the construction in Fig. 15(b). If k is
not too low the chord from the origin will intersect $T(R)$ in three places, at
which $p = 2k$, as shown in Fig. 17 (left). Now imagine k to decrease
gradually until two of the intersection points of the chord with the curve
come together at a tangency point. This value of R locates a pressure pit as
shown in Fig. 17 (right). □

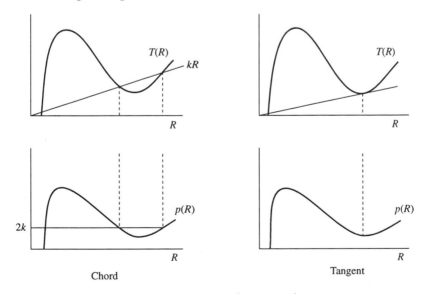

Fig. 17. Pressure pit construction

It can be seen that the $p(R)$ graphs in Figs 16 and 17 are steep where they begin on the left. This answers the question in Section 1(a). A party balloon is hard to blow up at first, because the initial steep slope of $p(R)$ means a stiff response. In other words, a large increase in p implies only a small increase in R.

11. Comparison of tension and pressure values

Before we can compare numerical values of T and p we need to express them in the same units. The dimensions of T and p are force per unit length and force per unit area respectively, so we need to divide T by some constant quantity having dimensions of length before we can compare like with like. We can use the natural radius r of the balloon to define

$$t = \frac{T}{r} \quad \text{and} \quad s = \frac{R}{r} \quad \text{so that} \quad p = \frac{2t}{s} \tag{11}$$

from $p = 2T/R$. Here s is the stretch introduced in Section 9. As r is fixed, the graph $T(R)$ converts into another, but qualitatively unchanged, graph $t(s)$. With $p = 2t/s$ from equation (11), this defines $p(s)$, which will be a qualitatively unchanged version of $p(T)$. We are now in a position to draw the functions $t(s)$ and $p(s)$ on the same *vertical* as well as horizontal scales, as shown in Fig. 18 with the tangent construction to relate the pressure peaks and pit.

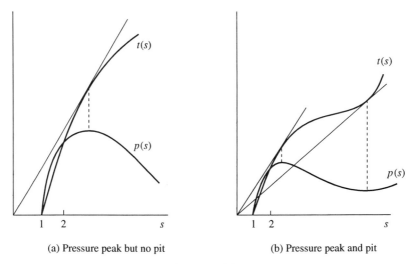

(a) Pressure peak but no pit   (b) Pressure peak and pit

Fig. 18.   Examples of $t(s)$ and $p(s)$ with tangent constructions

Theorem 8

The two graphs of $t(s)$ and $p(s)$ intersect at $s = 1$ and $s = 2$. Elsewhere $p > t$ in the range $1 < s < 2$, and $p < t$ in the range $s > 2$.

Proof

The intersection occurs where $p = t$ at the same s. From $p = 2t/s$ in (11), this requires $t(1 - 2/s) = 0$. This is satisfied by $t = 0 = p$ at $s = 1$ (i.e. at the start of the graphs), but also at $s = 2$. Put otherwise, $t(2) = p(2) > 0$.

The graphs are only defined for $s \geq 1$. We have

$$p = \frac{2t}{s} < t \quad \text{if} \quad s > 2, \quad \text{and} \quad p > t \quad \text{if} \quad 1 < s < 2. \qquad \square$$

Figure 18 shows the case in which the intersection of $t(s)$ and $p(s)$ at $s = 2$ occurs before the pressure peak. This might be the most likely case physically, because rubber can stretch to several times its natural length; but other cases are possible mathematically (see question 3 of Worksheet 2).

Worksheet 2

1. Pull a straight elastic band, and measure the maximum stretch which can be achieved before rupture.
2. Suppose that the type of rubber used in a spherical balloon is such that the dependence of T upon R is one of the types shown here.

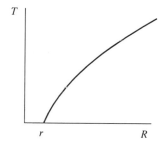
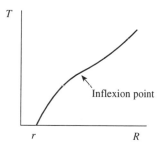

Prove that when R is the equilibrium radius which satisfies $R > r$, $p < 2T/r$. There is no local peak of $T(R)$ here, in contrast to Fig. 15. The slope is always positive, and on the right has a minimum at the inflexion point. The tangent there does not intersect the positive R-axis.

Justify and apply chord and tangent constructions like those of Fig. 16 to prove the existence of, and locate, pressure peaks in both cases. Repeat, following Fig. 17, for a pressure pit in the case on the right.

3. Sketch the versions of Fig. 18(a) in which the intersections of $t(s)$ and $p(s)$ at $s = 2$ occur (a) at the pressure peak, (b) beyond it and before the tension peak, (c) at the tension peak and (d) after the tension peak.
4. Recall the hypothesis in Section 5 that the tension T in a soap film is the same across every direction in the film at the considered point, and that equation (8) holds in equilibrium. Hence prove that a soap bubble in equilibrium cannot have the shape of a circular cylinder with hemispherical ends such as is adopted by some rubber balloons. Does this accord with observation?

5. This question should only be attempted by those who have met the idea of differentiating a function. Suppose that two functions $p(R)$ and $T(R)$ are related by $p = mT/R$, where m is a constant ($m = 1$ for circular arcs and $m = 2$ for spherical patches).
   Prove that

$$\frac{\mathrm{d}p}{\mathrm{d}R} = \frac{m}{R}\left(\frac{\mathrm{d}T}{\mathrm{d}R} - \frac{T}{R}\right).$$

Show that a straight line from the origin is a tangent to $T(R)$ where $\mathrm{d}T/\mathrm{d}R = T/R$, and prove that this implies a stationary point of $p(R)$ in the sense that $\mathrm{d}p/\mathrm{d}R = 0$.

12. Spherical and aspherical expansion

When a rubber balloon is filled with air, it first takes up the slack until it achieves its *natural shape*. The membrane tension and excess pressure are then both zero, but about to increase from that.

Suppose the natural shape is spherical, with radius r. Figures 15 to 18 show how p and T then depend on the size, as measured by R or s. After the initial stiff phase, remarked upon at the end of Section 10, the inflation becomes markedly easier to achieve as the radius grows to values near the pressure peak, at which there would be an instability of the system if the pressure were assigned in value. In practice your lungs detect the incipient instability and react by *relaxing* the pressure and assigning the radius, so that the balloon continues to expand.

It has been found experimentally by Harold Alexander (1971) and myself (1976, 1990) that, with *certain* types of rubber, a balloon which is spherical in the first phase of rising $T(R)$ and $p(R)$ graphs subsequently becomes aspherical. These features are shown in Figs. 19 and 20 for a meteorological balloon, i.e. for one of the types used to send instruments aloft into the atmosphere. The manufacturers' seam serves as the equator in Fig. 19, and its new position has been drawn in Fig. 20 for clarity.

Alan Needleman (1977) and David Haughton (1980), in analyses of a spherical membrane with and without geometrical imperfections respectively, verified my 1976 conjecture that a *closed loop* graph like that shown in Fig. 21 describes how that asphericity can develop and disappear, even in a balloon with *no* geometrical imperfections. This is the very striking result mentioned in Section 1(b). Alexander (1971) gives a photograph showing experimentally the recovered sphericity at the larger size.

Summarizing, for certain rubbers the spherical shape of a geometrical perfect balloon is *stable* before the peak and after the pit in the tension curve, and it is *unstable* between them where the aspherical equilibrium is stable. *Branching* of the stable equilibrium solution from spherical to aspherical and back occurs at the tension peak and pit respectively, and therefore strictly between the pressure peak and pit, as shown schematically in the three-dimensional graph of Fig. 21.

Fig. 19. Spherical rubber balloon in the first expansion phase

Fig. 20. Asphericity developed from a spherical balloon

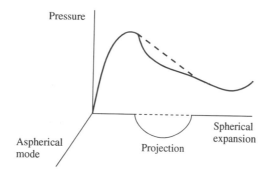

Fig. 21. Closed loop branching to aspherical balloon

13. Cylindrical balloon

When a cylindrical balloon is inflated, the sequence of events noted in Section 1(c) occurs. More control is achieved if the inflation is carried out with a balloon pump, obtainable from toy shops, rather than with your lungs.

The inflation is then controlled by adding assigned volumes of air, leaving the pressure to find its own level. It has been suggested by E. Chater and J. W. Hutchinson (1984) that the pressure/volume relation will be as shown in Fig. 22 (for spherical balloons the volume $= 4\pi R^3/3$, so volume can be replaced by radius R as previously, but not so in the present problem).

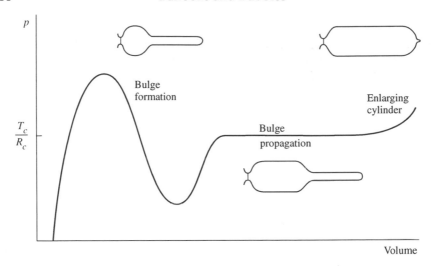

Fig. 22. Expansion of cylindrical balloon

The balloon is hard to inflate at first, as the initial high slope of the graph indicates, and as your lungs would testify. The pseudo-spherical bulge forms near the pressure peak and it has become established near the pressure pit. Just as the pressure begins to rise again, the increasing resistance which a single sphere could offer (see Fig. 15(b)) is balanced by a weakness in the saddle shaped part of the balloon which joins the pseudo-sphere to the remaining cylinder. This weakness is reminiscent of the fact that a saddle shaped soap film can be in equilibrium under zero pressure. Put otherwise, the longitudinal component T_l (say) of membrane tension in the saddle shaped part will vary with position in rubber, but acts outwards to *assist* instead of oppose the pressure, as shown in Fig. 23. This allows the bulge to propagate quasi-statically under a constant pressure equal to

$$\frac{T_c}{R_c} = \frac{2T_s}{R_s},$$

(12)

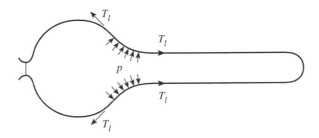

Fig. 23. Longitudinal forces on a saddle shaped patch of balloon

because the pressure must be the same everywhere, in equilibrium. Here R_c and T_c are the radius and hoop tension in the cylinder, R_s and T_s are the radius and tension in the spherical patch, and with $R_c < R_s$ so that $T_c < 2T_s$ (compare question 4 of Worksheet 2).

Solutions to Worksheet 1

1. $120°$.
2. Downward force $2T =$ upward force $2Rp$, so $p = T/R$.
3. There is no horizontal force on the chord, and by symmetry the internal pressures exert no net horizontal force on the curve either. Therefore, if the whole boundary is to be in equilibrium, and since no other forces are acting on the boundary as a whole, the vertically upward resultant of the pressures on the curve must be equal to the vertically downward resultant of the pressures on the chord, which is $2cp$.
4. Downward force $2\pi RT =$ upward force $\pi R^2 p$, so $p = 2T/R$.
5. A similar argument to that which solves question 3 applies, except that the flat part of the boundary is now circular with area πc^2, so the downward resultant pressure on it is $\pi c^2 p$. Equilibrium requires that the upward force on the curved part of the cap has the same value.
6. The circumferences has length $2\pi c$. Horizontal tension components equilibrate themselves, by symmetry. Vertical tension components are Tc/R per unit length, by similar triangles. The resultant vertical tension force is therefore $2\pi c^2 T/R$.

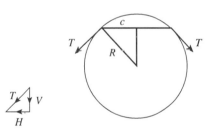

7. $\pi c^2 p = 2\pi c^2 T/R$ implies $p = 2T/R$.
8.

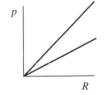

Straight lines	Hyperbolae	Equilibrium surface
R=constant	T=constant	

Solutions to Worksheet 2

1. A stretch value of 5 should be readily achieved.

2. $p = \dfrac{2T}{R} = \dfrac{2Tr}{rR} < \dfrac{2T}{r}$ because $\dfrac{r}{R} < 1$.

3.

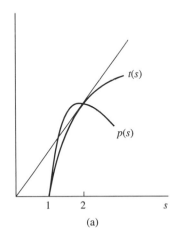

(a)

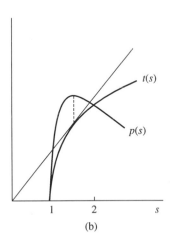

(b)

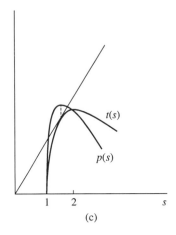

(c)

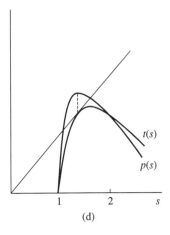

(d)

4. A circular cylinder of radius R has principal radii of curvatures $R_1 = R$ and $R_2 = \infty$, so the equilibrium pressure would have to be $p = T/R$ according to $p = T(1/R_1 + 1/R_2)$. As Theorem 2 shows, the formula is consistent with T being the loop stress within the circular cross-section perpendicular to the axis of the cylinder. But the tension in the cylinder in its axial direction, such as would have to balance the effect of the excess pressure exerted on the hemispherical ends, would have to be T such that $p = 2T/R$ given by

Theorem 3. This inconsistency, between $p = T/R$ and $p = 2T/R$, shows that the membrane tension cannot be the same in every direction within the membrane. The latter basic hypothesis for a soap film (see Section 5) is therefore contradicted, so a soap film cannot form such a closed cylinder in equilibrium with hemispherical ends. It seems never to have been observed.

5. $$\frac{dp}{dR} = \frac{m}{R}\frac{dT}{dR} - \frac{mT}{R^2} = \frac{m}{R}\left(\frac{dT}{dR} - \frac{T}{R}\right).$$

In the R, T plane the gradient of a straight line from the origin has the constant value T/R. The local gradient of the function $T(R)$ is dT/dR. Therefore they are equal at points where $dT/dR = T/R$. At such points $dp/dR = 0$ from the above equation, whatever $m(\neq 0)$ may be. Hence the implied function $p(R)$ is stationary there. Until more information is given, the stationary point may be a local minimum, or a local maximum, or a local horizontal inflexion point. You may sketch these possibilities.

Acknowledgements

I am indebted to Mr. Simon Johnson, of the Reading University Photographic Service, who took all the photographs at my behest. Figure 11 was first published in my 1993 article referenced below, and the apparatus there was kindly provided by Dr. G. W. Green of the Reading University Physics Department. Figures 19 and 20 were published in my 1976 and 1990 articles, and the meteorological balloon there was given to me by Dr. J. R. Milford of the Reading University Meteorology Department. Figure 8 appeared first on the cover of my book (1987). Figures 9 and 10 appeared in the *University of Reading Centenary Year Annual Report 1991–92.*

References

H. Alexander, 'Tensile instability of initially spherical balloons', *International Journal of Engineering Science*, **9** (1971) 151–162

E. Chater and J. W. Hutchinson, 'On the propagation of bulges and buckles', *Journal of Applied Mathematics*, **51** (1984) 269–277

D. M. Haughton, 'Post-bifurcation of perfect and imperfect spherical elastic membranes', *International Journal of Solids and Structures*, **16** (1980) 1123–1133

A. Needleman, 'Inflation of spherical rubber balloons', *International Journal of Solids and Structures*, **13** (1977) 409–421

M. J. Sewell, 'Some mechanical examples of catastrophe theory', *Bulletin of the Institute of Mathematics and its Applications*, **12** (1976) 163–172

M. J. Sewell, *Maximum and minimum principles. A unified approach, with applications.* (Cambridge University Press, 1987)

M. J. Sewell, 'Demonstrations of buckling by bifurcation and snapping' *Elasticity: mathematical methods and applications*. The Ian N. Sneddon 70th birthday volume (eds. G. Eason and R. W. Ogden) (Ellis Horwood, Chichester, 1990) 315–345

M. J. Sewell, 'Mathematics masterclasses for young people – objectives and scope', *Bulletin of the Institute of Mathematics and its Applications*, **29** (1993) 34–41

Further Reading

C. V. Boys, *Soap bubbles: their colours and the forces which mould them* (Dover, New York, 1991)

This is a new reprint of the classic text about demonstrations of soap films and related subjects, first published in 1890.
It was reviewed in *New Scientist* on 24 August 1991 – 'A masterly exposition, it will delight ... both young and old.'

C. Eisenberg, *The science of soap films and soap bubbles* (Tieto Ltd, Clevedon, 1978)

This account includes many colour photographs of soap films in different configurations.

9   Dynamics of Dinosaurs

by MICHAEL SEWELL

Department of Mathematics, University of Reading

1. Introduction to some dinosaurs

We begin with some brief biological information, to make this account self-contained and to serve as a background for the subsequent mathematics.

Dinosaurs were a particular subclass of reptiles. Many were essentially land animals, and included the largest such which have ever lived. The following are examples.

Diplodocus could be 27 metres long, which is $1\frac{1}{2}$ times longer than a cricket pitch, and with a head about the size of a rhinocero's head. It is shown in Fig. 1, fielding for the Reading University Academic Staff Cricket Club at mid-on (approximately), a position sometimes selected by the captain for Senior Lecturers whose fleetness of foot has become suspect. Its skeleton dominates the entrance hall of The Natural History Museum in London.

Fig. 1.   *Diplodocus*

Brachiosaurus grew to 13 metres tall, and therefore it could have looked over a three-storey house as shown in Fig. 2. The house is 9 metres high. Fossils of *Brachiosaurus* have been found in, for example, Tanzania and Colorado. A photograph of its skeleton in the Palaentological Museum of Humboldt University, Berlin was published by Halstead and Halstead (1981). For comparison, a giraffe grows to only $5\frac{1}{2}$ metres. *Brachiosaurus* and *Diplodocus* belonged to a leaf-eating group of dinosaurs called sauropods, which also included *Apatosaurus* (formerly called *Brontosaurus*). These sauropods were living around

Fig. 2. *Brachiosaurus*

140 million years ago (140M in the notation for past time which we shall use in this account). They walked on all fours, with elephant-like fore legs which were of little use for handling things. Not all dinosaurs were so large, although many were very big.

It is desirable to be aware that not all dinosaurs were contemporary with each other. Each dinosaur species lived for only part, and sometimes only a small part, of the so-called Mesozoic geological era (230M to 65M). Therefore they did not all compete with each other. The Rockies, the Andes and the Alps are all more recent than the dinosaurs. The older Appalachians have had more time to be eroded and worn down. It is helpful to construct a time line like that in Fig. 3. The reasons for the disappearance of the dinosaurs 65 million years ago is still a matter of debate, but a recent theory attributes it to a climate change caused by the impact of a very large meteor in the Gulf of Mexico.

Compsognathus (140M) was one of the smallest members ($1\frac{1}{2}$ metres long) of a flesh-eating group of dinosaurs called theropods, which also included the late and much larger *Tyrannosaurus rex* (70M), shown shadowing the University of Reading administration building in Fig. 4. This had 15 cm teeth which were serrated like a steak knife. It was the biggest theropod known until the 1993 find of 70% of the bones of *Giganotosaurus carolinii* (97M) in Patagonia was reported by Coria and Salgado (1995). The latter was also a biped, but it was $12\frac{1}{2}$ metres long. It head was $1\frac{1}{2}$ metres long.

Other plant eaters included *Psittacosaurus* (90M), which grew only to about 2 metres, and had a beak-like rather than a toothed jaw; and the much bigger

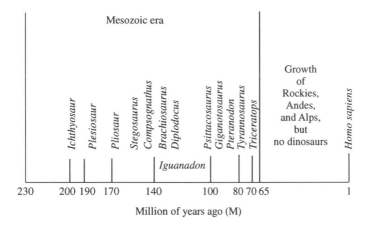

Fig. 3. Time line showing presence of dinosaurs

Iguanodon (140M to 100M) shown in Fig. 5, of which there is a complete skeleton in the University Museum, Oxford. These walked on their hind legs, and had much smaller fore legs, like *Compsognathus* and *Tyrannosaurus*.

Plant eaters which walked on four legs included the horned dinosaurs such as *Stegosaurus* (140M, Fig. 6), which had plates on its back and spines on its tail, and *Triceratops* (65M, Fig. 7). They were, respectively, about two and six times as massive as a modern rhinoceros. *Triceratops* probably used its three horns to

Fig. 4. *Tyrannosaurus*

Fig. 5. *Iguanodon*

defend itself against *Tyrannosaurus*, in North America, but *Stegosaurus* was extinct long before either appeared.

Fig. 6. *Stegosaurus*

Fig. 7. *Triceratops*

Pterosaurs were winged reptiles contemporary with dinosaurs, such as *Pteranodon* (80M, Fig. 8). It had a wing span of 7 metres, bigger than some hang gliders, and twice as much as the wandering albatross, which has the biggest span of modern birds.

Fig. 8. *Pteranodon*

Fig. 9. *Plesiosaur*

Giant marine reptiles of the Mesozoic era included *Ichthyosaur* (200M to 190M, Fig. 11), which could be 8 metres long, and therefore comparable in size to the modern killer whale or great white shark; *Plesiosaur* (190M to 170M, Fig. 9) which reached 14 metres in length; and *Pliosaur* (165M, Fig. 10) which could be 10 metres long.

It is of the order of 1M since *Homo sapiens* appeared on the scene, perhaps less than 0.005M (5000 years) since he began to understand his environment in any modern sense, and less than 0.0002M (200 years) since he developed the capacity to destroy it.

Fig. 10. *Pliosaur*

2. Weighing dinosaurs

How can we weigh a dinosaur? There are no live specimens available, nor even dead ones still possessing their flesh and blood as well as bones. Even if there were, we have no scales large enough to hold even a medium-sized dinosaur. Big weighbridges like those used for lorries are available in zoos and cattle markets, and some are portable up to a point, but these are not convenient for big animals in the wild. It would be unreliable just to weigh fossilized bones, which have changed their chemical constitution by mineralization and loss of water. We have to weigh dinosaurs indirectly, and I shall describe two methods. First we need some definitions.

The *mass* of a body is the measure of intrinsic 'amount of matter' in the body. It is an intuitive concept. Common units of mass are the gram and the kilogram = 1000 grams. An astronaut who took a kilogram bag of sugar from the Earth to the Moon would find that it contained the same 'amount' of sugar when it arrived.

The *acceleration due to gravity* at a particular place is the acceleration (= rate of change of velocity) which *every* body would experience, in free fall, because of the pull of gravity. It has the same value whatever the mass of the body. This acceleration is denoted by g, and it is often measured in metres per second per second. It is an experimental fact that the value of g depends on the place, but not on the body. A feather and a piece of lead would fall side by side in a vacuum. This fact is masked when they fall in air, by the different aerodynamic resistance to the different shapes. On the Earth's surface $g = 9.81$ metres per second per second approximately (with slight variation depending on the precise location), and on the Moon's surface it is about $\frac{1}{6}$ of that value, i.e. about 1.63 metres per second per second. For rough calculations later in which we are mainly interested in getting the order of magnitude right, it will be adequate and convenient to take $g = 10$ metres per second per second.

The *weight* of a body is the force which the local gravity exerts on the body. Force is another intuitive concept. If w denotes weight and m denotes mass, then Newton's Second Law of Motion (published in 1687) tells us that

$$w = mg \qquad (1)$$

If m is measured in kilograms and g in metres per second per second, then weight is measured in 'newtons'.

Scales and spring balances all measure the weight of a body. This depends on the place, because g does so. Thus 1 kilogram of sugar weighs $1 \times 9.81 = 9.81$ newtons on Earth, but only $1 \times 1.63 = 1.63$ newtons on the Moon. This difference illustrates why an astronaut can jump higher on the Moon than the Earth, and why it may be more convenient to hop on the Moon rather than walk (at least for legs which have been designed to work on Earth).

The masses of different bodies can be compared by measuring their weights at the same place, or at places having the same value of g. Scales are always marked in units of mass, such as kilograms, rather than of the weight which they actually measure, such as newtons, on the assumption that the variation of g on the Earth's surface is negligible, so that g can be divided out as described below.

Subsequently we shall always take $g = 10$ metres per second per second for the reason given earlier. A man who weighs 700 newtons on Earth will have a mass of

$$m = \frac{w}{g} = \frac{700}{10} = 70 \text{ kilograms},$$

and is thus 70 times more massive than a 1 kilogram bag of sugar. For comparison, the approximate masses of some other modern animals are, in kilograms:

black rhinoceros	1170
hippopotamus	2520
African elephant	5450
blue whale	91 000

Dissection can be used to obtain some of these measurements on scales, i.e. directly, but for dinosaurs it cannot, as we have said. So what are the two indirect methods of weighing dinosaurs?

The first method uses *Archimedes' Principle*. This states that, when any object is wholly immersed in water, the water exerts an *upward* force on it which, when the water is still, is equal to the weight of as much water as would have the same *volume* of the body. Archimedes' Principle may be justified by the argument that, if the object were not there, the water would be, and in equilibrium. Archimedes lived in Syracuse, in Sicily, from 287BC to 212BC.

Archimedes' Principle is also called the *principle of buoyancy*, and it may be summarized as: *the upthrust is equal to the weight of water displaced*. This upthrust, or force of buoyancy, is what helps animals to float. It there is no attempt to swim, there are only two forces on an immersed animal, as shown in Fig. 11, namely the force of buoyancy b upwards, and the weight w of the animal downwards. The animal will

sink if $w > b$,   rise if $b > w$,   float if $b = w$.

Marine animals can float at any chosen level by controlling their volume (and hence b) via the air in their lungs. The reader will have experienced upthrust when trying to push an air-filled ball under water.

The volume of an awkward shape, when it is made of material which cannot be damaged by water, can be measured by immersing it in a vessel which is already filled to the brim with water, catching what is spilt, and measuring that volume in a measuring beaker. The method uses the fact that water is incompressible.

Alexander (1989) proposed an alternative method of measuring the volume of dinosaurs, by applying Archimedes' Principle to accurate plastic models of them, as follows. Suspend a scale model from one arm of a beam balance, over a

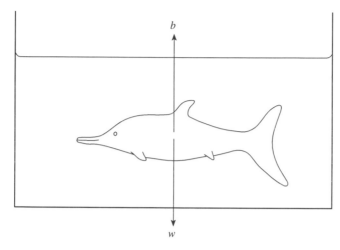

Fig. 11. *Icthyosaur* illustrating Archimedes' Principle

beaker, as shown schematically in Fig. 12. Weigh the model, by finding what weight is required to make the beam horizontal. Immerse the model by filling the beaker with water as necessary. Note what weight b (say) needs to be removed from the weights pan to make the beam horizontal again.

Theorem 1

b = weight of water required to occupy the volume of the model.

Proof
The two arms of the beam have equal length. Therefore the same weights s (say) are supported on each side at the start, and the same weights e (say) on each side at the end. So if $b = s - e$ is the weight removed from the pan, then $e = s - b$. Therefore b is the force of buoyancy, because

    end weight in pan = weight of model (down) − force of buoyancy (up).

Applying Archimedes' Principle then establishes the Theorem.   □

The following practical details are helpful to know when performing the weighing.

1. The model may float on the surface of the water if its density is near to that of water, as may be the case for a plastic model. To avoid this, a small metal counterweight needs to be attached, and immersed from the outset, i.e. before the model itself is immersed.
2. A standard-sized beam balance may have room only for a relatively small beaker, which implies that it will be easier to measure the upthrust b on a relatively small dinosaur model.
3. A bridge needs to be built over one scale pan, to carry the beaker.

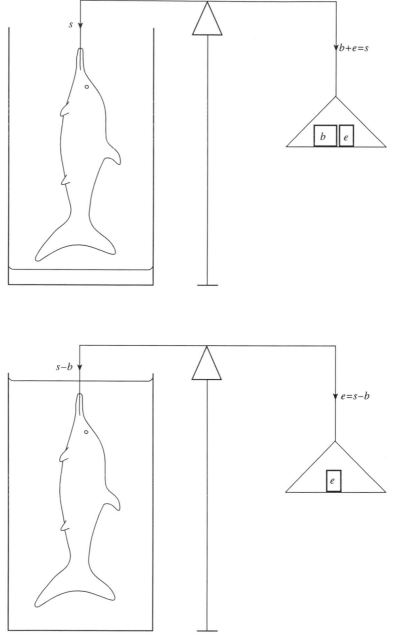

Fig. 12. Upthrust b = weight of water displaced

4. The balancing may be sensitive, e.g. to within a grain or two of rice which I find convenient to use at the last stages. The rice itself can be weighed on electronic scales, for example.

Theorem 2

If the weighing shows that b grams is the mass of water required to occupy the volume of the dinosaur model, and if it is an accurate model on a scale of $1 : l$ by length, then the mass of the corresponding real dinosaur was

$$\frac{l^3 b}{1000} \text{ kilograms.} \tag{2}$$

Proof
Density is defined as mass per unit volume, and the density of water at room temperature varies little from its maximum value of 1 gram per cubic centimetre at 4°C.

The volume of the dinosaur model is therefore b cc as determined via Archimedes' Principle; or it may have been so determined by direct displacement as described above. Either way, the volume of the full-sized dinosaur would have been $l^3 b$ cc. This is because, when length is scaled up by the factor l in every direction, area will be scaled up by a factor l^2, and volume by l^3.

Now most animals either just float or just sink in water, so it is reasonable to assume that the density of dinosaurs too was about equal to that of water. Therefore the mass of the real dinosaur was $l^3 b$ gm, or $l^3 b / 1000$ kg. □

The Natural History Museum in London sells good quality hard plastic models of at least 20 dinosaurs, which are on a scale of the order of $1 : 40$ by length. If b cc is the volume of such a model, the mass of the real dinosaur would have been

$$64 b \text{ kg} \tag{3}$$

according to the formula in (2).

I have found that their model of *Glyptodon* is a convenient size for the weighing experiment, and the formula just stated then gives a mass of 2560 kg for the real animal. The real *Glyptodon* was a tortoise-shaped animal at least 2 metres long, and in fact not a true dinosaur, because it was contemporary with the woolly mammoths only one million years ago.

The masses of some real dinosaurs quoted by Alexander (1989) to have been determined on the basis just described were, in kilograms:

Brachiosaurus	46 600
Diplodocus	18 500
Tyrannosaurus	7400
Triceratops	6100
Iguanodon	5400
Stegosaurus	3100

Other methods give different values, and there may be discrepancies either way which are not small. However, these figures do give ideas of orders of magnitude, e.g. for comparison with modern animals.

The second indirect method of weighing dinosaurs does not require knowledge of the complete animal, such as a reliable model or a complete skeleton. Only leg bones are required, and major leg bones are often well preserved in otherwise incomplete fossils. It seems to have been the case that dinosaurs stood and moved with the legs underneath them, and not splayed out to each side like other reptiles such as crocodiles. The method is based on these facts.

Anderson *et al.* (1985) measured the circumferences of the humerus and femur (the upper bones in the fore leg and hind leg respectively) of several dinosaur fossils with, for comparison, a number of modern mammals. The measurements were made half-way along, where the bone is narrowest at its 'waist'. They then proposed the following empirical formulae relating the mass *m*, measured in kilograms, of the whole animal to those circumferences, measuring in millimetres.

For bipeds

$$m = aC^b \quad \text{where} \quad a = 0.00016, \, b = 2.73 \tag{4}$$

and C is the circumference of the humerus. This formula underestimates the masses of kangeroos and overestimates the masses of ostriches. For comparison, a parabola has the formula $m = aC^2$ with any constant $a \neq 0$. The flight of a projectile or golf ball would be along a parabola if there were no air resistance.

For quadrupeds it is necessary to even out errors arising from the fact that different quadrupeds support different fractions of their total weight on fore legs and hind legs. This is done by using the *sum*, *c* (say), of the circumferences of one humerus and one femur. For *Brachiosaurus* the two measurements are similar (654 mm and 730 mm) but not so for the *Apatosaurus*, which evidently carried most of its weight on its hind legs. The formula

$$m = ac^b \quad \text{where} \quad a = 0.000084, \, b = 2.73 \tag{5}$$

is then proposed. This matches the data for a number of modern quadruped mammals. When it is used for dinosaurs, it gives estimates for their masses which are somewhat lower than those obtained by the Archimedes method quoted above.

Thus there is some uncertainty about details, but dinosaur masses found by this second method probably have the right order of magnitude.

Worksheet 1

1. The mass of a real dinosaur is estimated by the Archimedes method to be $m = 64b$ kg, where b cc is the volume of the 1:40 scale model.

   Work out the real masses from model volumes of $b = 290$ cc for *Diplodocus* and $b = 95$ cc for *Triceratops*.

   What model volumes would imply real masses of 7400 kg for *Tyrannosaurus* and 3100 kg for *Stegosaurus*?

   What does the formula $m = 64b$ become if the length scale factor is 1:45 instead of 1:40? What percentage increase does this imply for the mass of *Diplodocus*?

2. Complete the following table of values for the parabola $m = 0.25c^2$.

c	-4	-3	-2	-1	0	1	2	3	4
m									

Plot the curve. Repeat for $m = 0.5c^2$ and $m = 0.1c^2$.

3. Sketch the circular cross-section of a *Brachiosaurus* leg bone whose circumference is 70 cm.

4. Suppose that the mass m kg and circumference c mm of humerus + femur for quadrupedal dinosaurs are related by

$$m = ac^b \quad \text{where} \quad a = 0.000\,084, b = 2.73.$$

Use the x^y button on your calculator to find m from the following values of c, for *Diplodocus* (1130 mm), *Triceratops* (750 mm) and *Stegosaurus* (580 mm). Compare your results with values obtained by using $b = 2$ and then $b = 3$ as supposed approximations to $b = 2.73$. Are they good approximations?

5. For bipedal dinosaurs with femur circumference C mm, the mass m kg is

$$m = aC^b \quad \text{where} \quad a = 0.00016, b = 2.73.$$

Find m for *Tyrannosaurus* having $C = 640$ mm, for *Iguanodon* having $C = 570$ mm.

6. The theories of exponents and of logarithms tell us that if a set of data relating values of m and c satisfies the equation

$$m = ac^b \tag{A}$$

for some constant values of a and b, then the logarithms satisfy

$$\log m = \log a + b \log c. \tag{B}$$

Unless, $b = 1$, equation (A) is a curve in the c, m plane. It is a parabola if $b = 2$.

If $b = 1$, prove that equation (A) is a straight line which has gradient a and which passes through the origin $c = 0$, $m = 0$. Sketch it for some selected values of a, say $a = 1$ and $a = 2$. (The *gradient* of a straight line is defined to be the vertical step divided by the horizontal step between any pair of points on the line).

For any b, write $x = \log c$ and $y = \log m$, and hence prove by using a sketch that equation (B) is a straight line in (x, y) space which has gradient b and which passes through the y-axis at height $\log a$.

This property is the basis of the so-called log-log plot method of determining a and b when empirical scientific data is suspected of satisfying a curve of type (A) for unknown constants a and b. Logarithms of the data are plotted in x, y space, and a straight line of 'best fit' is selected. Its gradient b and intercept $\log a$ are then measured.

This is how the values $b = 2.73$ and $a = 0.000\,084$ or $0.000\,16$ were selected for quadrupedal and bipedal dinosaurs respectively. Your calculator may have logarithms to base 10 (often denoted by log) or to base $e = 2.718$ (often denoted by ln). Either will do.

3. Dinosaur footprints

What is the chance of a dinosaur footprint being preserved and found? Low perhaps, but not zero. Millions of dinosaurs made footprints every day, for millions of years. The search for, and interpretation of, dinosaur tracks has become a subject of study in itself, and books have been written about it, for example by Lockley (1991).

Figure 13 (from Alexander (1989) who cites Thulborn and Wade (1984)) shows some footprints from 100M found in Queensland. The biggest are 64 cm long, and were made by a bipedal dinosaur. It may have been a theropod, because of apparent claw marks at the end of at least the middle toe. The size suggests something of about 5000 kg, perhaps a smaller precursor of *Tyrannosaurus*. The other footprints were made by chicken-sized dinosaurs (not

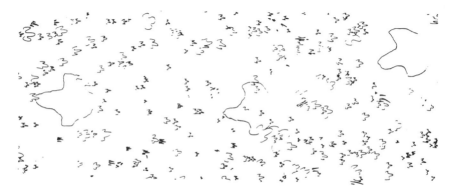

Fig. 13.   Bipedal footprints from Winton, Queensland

by birds, whose fossils are rare in contemporary rocks) going the opposite way, perhaps in trepidation.

Some footprints found in Texas are shown in Fig. 14 (also from Alexander (1989) who cites Bird (1944)). They include (stippled) those of a quadruped-like *Apatosaurus* with large oval hind footprints about 76 cm long, and smaller fore footprints, suggesting a mass of about 25 000 kg, or 25 tonnes.

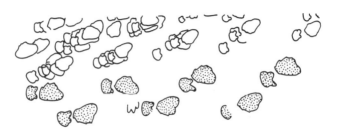

Fig. 14.   Quadrupedal footprints from Texas

Tracks of dinosaur tails are not usually found, and tails may *not* have dragged along the ground. In general, modern animals do not drag their tails; it would be hard on the tails. Instead it seems likely that tails were used either for balance, as a counterweight such as is illustrated by the kangaroo, or as a weapon, like a whip or club. *Stegosaurus* had spikes on its tail up to 50 cm long.

What can we deduce from dinosaur footprints? In many examples like the ones in Figs. 13 and 14 the two lines of left and right prints are rather close together, indicating that those dinosaurs had walked with their feet *under* their body, like modern birds and mammals, and not splayed out to the side like crocodiles. The shape of dinosaur leg bones indicates the same thing. This deduction led to the second method of estimating the weight of dinosaurs described in Section 2.

Did dinosaurs get bogged down? Of course they did, and perhaps quite often; but can we make some quantitative calculations about this question? The answer depends not only on the size and shape of the dinosaur, but also on the type of ground. We need to define some new concepts.

Stress is defined as force per unit area. When a bipedal dinosaur whose weight is w newtons (say) is standing symmetrically, the compressive stress s (say) acting over a foot or a cross-section of leg whose area is $\frac{1}{2}A$ square centimetres is $\frac{1}{2}w/\frac{1}{2}A$, i.e.

$$s = \frac{w}{A} \text{ newtons per square centimetre.} \tag{6}$$

Theorem 3

A big dinosaur, which was twice as long, wide and high as a similarly shaped small one with the same density, would exert twice the stress on the ground. It would therefore be twice as likely to sink in.

Proof

If each length is scaled by a factor l, areas will be scaled by a factor l^2, and volumes will be scaled by a factor l^3.

Weights would also be scaled by a factor l^3 if the two dinosaurs had the same density $d = m/v$, where $v =$ volume, because $w = mg = gdv$ is proportional to v with the common factor gd.

The stress $s = w/A$ transmitted by a foot to the ground would therefore be scaled by a factor $l^3/l^2 = l$.

Therefore, if $l = 2$ for example,

$$2 \times \text{length} \Rightarrow 4 \times \text{area} \Rightarrow 8 \times \text{volume} \Rightarrow 8 \times \text{weight},$$

and the stress on the ground would be doubled.

Ground stress, being force per unit area, is what controls the likelihood of sinking.   □

Whether sinking actually occurred would depend on the type of ground, because different types offer different support. The strength of wet clay depends on cohesion between particles, but that of dry sand depends on friction between particles.

The *yield stress* of a material is defined to be the particular value of stress which causes it to collapse. The strength of ground is measured by its yield stress. Ground will only support bodies, whether they be animals or buildings, whose weight exerts a stress upon them which is below the yield stress. Stresses at or above that value will cause the ground to give way and the body to sink. The longer legs of a big dinosaur may have given it some advantage over a small one in that situation.

Let y denote the yield stress of the ground, and A now denote the area over which weight is being applied. Thus for a standing animal, A is now the area of four feet for a quadruped and two for a biped. Empirically it is found that

$$y = aA^b \tag{7}$$

where a and b are constants whose value depends on the type of ground.

For clay $b = 0$, so that (because $A^0 = 1$)

$$y = a$$

is actually independent of area, but the value of a depends on the water content. For sand $b = \frac{1}{2}$ so that (note that $A^{1/2} = \sqrt{A}$)

$$y = aA^{1/2},$$

i.e. the yield stress depends on the contact area parabolically, as shown in Fig. 15, as well as on the constant a whose value will depend on the size and shape of the grains.

The stress exerted by an animal of weight w standing on an area A is w/A, so the ground is *safe* if $w/A < y$, i.e. if

$$w < Ay$$

but ready to yield if

$$w = Ay = aA^{b+1} = W \text{ (say).} \tag{8}$$

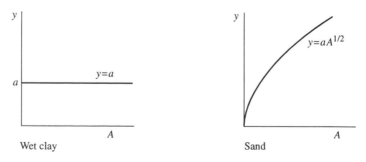

Fig. 15.   Dependence of yield stress y on contact area A

Thus we use W to denote the *maximum* weight of animal which can be supported on an area A of ground.

Theorem 4

Sand would support a big dinosaur just as well as a small one.

Proof
Using $b = \frac{1}{2}$, the maximum weight of dinosaur which can be supported on an area A of sand is
$$W = aA^{3/2}.$$
If the dinosaur is scaled up by a length factor $l > 1$, its feet area will be scaled by a factor l^2 and, by the argument in the proof of Theorem 3, its weight w will be scaled by a factor l^3. When it is standing on sand we see that
$$l^3 W = a(l^2 A)^{3/2},$$
i.e. the maximum supportable weight is scaled by exactly the same factor as the actual weight. This is another way of stating the Theorem. ☐

For example, $W = aA^{3/2}$ implies $8W = a(4A)^{3/2}$ when lengths are doubled ($l = 2$), because $8 = 2^3 = (4^{1/2})^3 = 4^{3/2}$.

Theorem 5

Clay would *not* support a big dinosaur as well as a small one.

Proof
Using $b = 0$, the maximum weight of dinosaur which could be supported on an area A of clay is
$$W = aA.$$
Therefore $l^2 W = a(l^2 A)$. This implies that, when the representative dinosaur length is scaled up by a factor $l > 1$, so that its feet area is scaled by a factor l^2, the maximum supportable weight W is *also* scaled by a factor l^2; whereas its true weight w will (as before, for example in Theorem 4) be scaled by a factor $l^3 > l^2$. In other words, the true weight of the larger dinosaur would exceed the maximum supportable weight, which proves the Theorem. ☐

For example, doubling the lengths ($l = 2$) quadruples the maximum supportable weight because $4W = a(4A)$, but the true weight is increased eight-fold.

	Mass m (kilograms)	Total foot area A (square metres)	Ground stress w/A (newtons per square metre)
Apatosaurus	35 000	1.2	290 000
Tyrannosaurus	7000	0.6	120 000
Iguanodon	5000	0.4	125 000
African elephant	4500	0.6	75 000
Domestic cattle	600	0.04	150 000
Human	70	0.035	20 000

The table, which is based upon information given by Alexander (1989), shows some approximate values of the stress exerted by different animals on the ground. Wet clay having yield stress $y = 100\,000$ newtons per square metre will support man and elephant, but not the others. If $y = 130\,000$ the two bipedal dinosaurs are safe, but not cattle. Cattle are safe if $y > 150\,000$, but *Apatosaurus* is not until $y > 290\,000$. The relative safety of elephants is because of their relatively large feet, whereas cattle are at high risk of sinking because of their small feet.

4. Speed of dinosaurs

How fast could a dinosaur move? Can we answer this question by measuring footprints and fossil bones?

Tracks of footprints reveal the *stride length* S (say), defined to be

S = distance from one footprint to the same

point on the next print of the same foot.

The stride length is longer in running than in walking, so speed is certainly related to stride length, but the evidence from modern animals is that this relation is different for different species. We do not know it for dinosaurs.

We want to make a prediction which will be valid, if not for all, then at least for a wide range of species, having different sizes, so that we can apply it to dinosaurs. It turns out that this may be achieved if we also suppose that speed is related to *leg length* L (say), defined to be

L = height of hip in normal standing.

Measure your own S and L. My S varies between 80 cm when I am walking thoughtfully, 96 cm when I am walking purposefully, and 160 cm when I am sprinting between wickets. Of course, in using S in this theory we shall presume it to be achieved with some normal physiological effort, and not artificially as, for example, in the soldier's slow march. My $L = 96$ cm when wearing shoes (I shall want to estimate my shod speed).

An appropriate *dimensionless* measure of leg action is *relative stride length* r (say), defined to be

$$\frac{\text{actual stride length}}{\text{leg length}}, \quad \text{i.e. } r = \frac{S}{L}. \tag{9}$$

This can be used to compare the performance of animals of different absolute sizes if we introduce the *principle of dynamical similarity*. This is exploited, for example, whenever small scale models are made to test a larger real-life situation. The principle is that the mechanical behaviour of the model will mimic that of the real thing if all lengths are scaled by one factor, all times by another factor, and all forces by a third factor, provided the factors are correctly chosen.

Ship laboratories use this principle. Hydraulics Research at Wallingford has a very large building with a model of the Severn estuary in it, to explore tidal flows. In the opposite sense, there are cases where a larger scale model is used to explore a tiny real-life situation: in physics, for example, ball bearings are projected at a stretched rubber membrane so that the latter can mimic gravitationally the electric potential distribution in an electron gun.

To apply the idea of dynamical similarity in our situation, where we wish to determine the true speed v (say) of an extinct animal, we define the *dimensionless speed u* (say) by

$$u = \frac{v}{(Lg)^{1/2}}. \tag{10}$$

In this fraction the numerator has the dimensions of distance/time. The denominator has the dimensions of $[\text{distance} \times \text{distance}/(\text{time})^2]^{1/2} = $ distance/time also. So the quotient u has no dimensions.

Alexander (1989) and his co-workers found that, for all of the widely different species of modern animal mentioned below, including bipeds and quadrupeds, experimental data lay reasonably near a particular straight line

$$r = au + b \tag{11}$$

in the u, r plane, where a and b are particular constants. The intercept on the vertical axis is b (the value of r where $u = 0$), and the gradient of the line is $a = (r - b)/u$. It is a piece of luck that the experimental relation between r and u is as simple as a straight line, but we shall see that there is a particular mathematical caution to be observed. This is that, after we have used equations (9) and (10) to rewrite (11) as

$$v = \frac{(Lg)^{1/2}}{a} \left(\frac{S}{L} - b \right), \tag{12}$$

the consequent prediction of the true speed v from the measurements S and L may be sensitive to the particular values of a and b which are thought to 'best fit' the data.

Measuring Alexander's chosen line shows it to have gradient $a = 1.38$ and intercept $b = 0.75$ as shown in Fig. 16. Apart from dogs, no species comes near

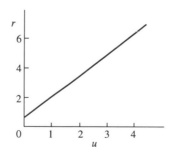

Fig. 16.   Alexander's line $r = 1.38u + 0.75$

to occupying the whole line, but they each occupy a different part of it as specified by the following ranges of u:

ostrich	$0.7 < u < 1$
dogs	$0.5 < u < 5$
sheep	$2.3 < u < 3.6$
camel	$0.3 < u < 2$

human	$1.8 < u < 2.8$
elephant	$0.8 < u < 1.1$
rhinoceros	$1 < u < 2.2$

Corresponding ranges of r can be calculated from equation (11).

The data fitted by Fig. 16 is for animals moving on hard ground not making footprints, whereas dinosaur stride lengths are measured on ground which was soft at the time the footprints were made. Alexander suggests, however, that the type of ground does not affect stride length, and therefore that Fig. 16 can also be used for dinosaurs.

To do that we may need, finally, a way of inferring leg length from *foot length* F (say). This is because, although foot length can be measured from the footprints, there may be no complete fossilized leg remaining with it. Measurement of skeletons which do exist, however, suggest that a general rule might be

$$L = 4F. \tag{13}$$

Measure your own foot length to compare with this rule. Mine is 33 cm wearing shoes, so my $L/F = 96/33 = 2.9$, which is significantly lower than 4.

Noting this discrepancy, but nevertheless combining the empirical formulae in equation (13) and Fig. 16 offers a particular version of equation (12) for estimating the true speed of dinosaurs from footprint measurements, namely

$$v = \frac{(4Fg)^{1/2}}{1.38}\left(\frac{S}{4F} - 0.75\right). \tag{14}$$

The following examples are from Figs. 13 and 14 (using 1 metre per second = 2.25 miles per hour).

	F cm	S cm	r	u	v mph
Large Queensland biped	64	331	1.3	0.4	4.5
Stippled Texas quadruped	76	279	0.92	0.12	1.51

These may be easy walking speeds for such large dinosaurs, but the biped speed would be very fast walking for a human. Zoologists observe that mammals and birds walk when their $u < 0.7$, and run when $u > 0.7$.

My own measurements, with $a = 1.38$ and $b = 0.75$ as in Fig. 16, imply

Gait	S cm	$r = S/L$	$u = \dfrac{r - 0.75}{1.38}$	v mph
Slow walk	80	0.83	0.06	0.42
Fast walk	96	1.00	0.18	1.25
Sprint	160	1.67	0.67	4.68

The human fast walking speed is known to be of the order of 4 mph, so these predictions are plainly much too low, perhaps by a factor of 3. It is important to ask why this is so. Note first, however, that the empirical rule (13) was not used (as it was for the dinosaur calculation) because L is known directly and $L = 2.9F$. It would have made matters much worse to have used equation (13), because $S/4F - 0.75 = 96/132 - 0.75 = -0.02$ is *negative* even for the fast walk.

Avoiding equation (13), therefore, the other sources of error are in the choice of the values of a and b in equation (12) to fit a straight line to the data. In fact, if a straight line had been sought to fit *only* the human data quoted by Alexander, a lower value of a (than 1.38) and a higher value of b (than 0.75) would have been chosen. The first change would work in the right direction, but the second change would work in the wrong direction. Evidently fitting data by straight lines is a delicate matter.

Measure your own S, F and L and examine how the consequent predictions of v are affected by different choices of a and b. Persuade your friends to do the same, and see if you can find the 'best fit' choices of a and b for humans.

Worksheet 2

1. If a leg bone has a circular cross-section with circumference C and area $\frac{1}{2}A$, prove that $\frac{1}{2}A = C^2/4\pi$.
   If it belongs to a bipedal dinosaur of weight w which is standing symmetrically, use the formula $s = w/A$ to show that the stress over that cross-section is $s = 2\pi w/C^2$.
   Use $w = mg$, with $g = 10$ metres per second per second approximately, to prove that $s = 63m/C^2$.
   For *Iguanodon* with $m = 5341$ kg and $C = 0.57$ metres, calculate s and specify its units. Repeat for *Tyrannosaurus* with $m = 7328$ kg and $C = 0.64$ metres.
2. If a quadrupedal dinosaur of weight w could stand symmetrically on all four legs, prove that the stress s over a leg cross-section whose area is $\frac{1}{4}A$ is $s = w/A$.
   Prove that if C is the circumference of such a cross-section, $s = \pi w/C^2$.
   Use this with $g - 10$ metres per second per second to complete the following table of values for s and specify its units.

	m kg	C cm	s
Brachiosaurus	46 600	70	
Diplodocus	18 500	57	
Triceratops	6100	37	
Stegosaurus	3100	29	

3. The real speed v of a ship having hull length L can be used to define a dimensionless speed $u = v/(Lg)^{1/2}$.

Find the dimensionless speed of a real ship 300 metres long which is travelling at 15 metres per second, using $g = 10$ metres per second per second.

What should be the real speed of a 5 metre model in order to achieve the same dimensionless speed?

Ship designers use the principle of dynamical similarity to avoid expensive mistakes in real ships by building small scale models and testing their performance in large tanks.

4. Use the fact that 8 kilometres = 5 miles to prove that 1 metre per second = 2.25 miles per hour.

5. Use $r = 1.38u + 0.75$ to calculate the ranges of r corresponding to the following ranges of u:

ostrich	$0.7 < u < 1$	human	$1.8 < u < 2.8$
dogs	$0.5 < u < 5$	sheep	$2.3 < u < 3.6$

6. Suppose that a dinosaur has $F = 40$ cm and $S = 180$ cm.
   Find its true speed v, in units to be specified, by using

$$v = \frac{(4Fg)^{1/2}}{1.38} \left(\frac{S}{4F} - 0.75 \right).$$

7. Measure your own foot length F, leg length L, and brisk walking stride length S. Calculate L/F. Assuming that your brisk walking pace is $v = 3$ mph, convert this to metres per second. Then, using consistent units and $g = 10$ metres per second per second, plot the straight line

$$a = \frac{(Lg)^{1/2}}{v} \left(\frac{S}{L} - b \right)$$

in a plane with a measured on the vertical axis and b on the horizontal axis. Measure your sprinting stride length S, and assuming that you can sprint 100 metres in 20 seconds, plot another such straight line on the same diagram. Find the values of a and b where the two straight lines intersect. Discuss the significance of these values for the human race.

5. Athletic animals

Large is lumbering, like an elephant. Neat is nimble, like a gazelle. A buffalo is somewhere in between. Can we construct a scale of athleticism, upon which to base predictions? Were those dinosaurs which were much larger than elephants also much less nimble?

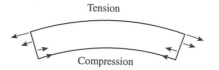

Tension

Compression

Fig. 17. Bent bar under tension and compression

Athletic performance depends on several factors, varying from the strength of bone and muscle to the efficiency of oxygen consumption. Different objectives have different prerequisites, as a comparison of sumo wrestlers and marathon runners shows. I refer again to Alexander (1989) for a many-sided account oriented towards dinosaurs, including the performance of marine and flying reptiles.

Here I note some remarks which he makes in relation to the *bending* of leg bones in land animals. A bar bent as in Fig. 17 may have tensile forces acting along the fibres on the convex side, and compressive forces along the fibres on the concave side, as suggested by the arrows. A force system like that can be equivalent in its static effect to those shown, as bending begins, on half the bar represented as a cantilever in Fig. 18. That is, simple bending theory shows that a fibre *in equilibrium* whose distance is x from the centre line carries a stress $s = -Bx/Z$, where $B = Fz$ is bending moment associated with a transverse force F applied at a distance z from the support. Here Z is a number called the moment of inertia which depends on the cross-sectional shape of the bar. It is about four times bigger for a so-called I-section beam than for a circular-section beam of the same area. The former is therefore stronger because, at the same place, the stress is less.

Suppose the reaction of the ground to the foot of *Tyrannosaurus* is a force with components P parallel to the tibia and R transverse to it, as shown in Fig. 19, which is based on Alexander (1989, Fig. 4.3). Let M be the force exerted by the calf muscles. The reactions between ankle and tibia are $M + P$ and R, as shown by Newton's Third Law, which states that action and reaction are equal and opposite. This combination of pure compression and bending induces a longitudinal stress in the fibres of the tibia, at a cross-section of area A which is distance z from the lower end, which is

$$-\frac{(M+P)}{A} - \frac{Rzx}{Z} \tag{15}$$

$$\text{Stress } s = -\frac{Bx}{Z} \qquad \text{Transverse force } F \qquad \text{Bending moment } B = Fz$$

Fig. 18. Statically equivalent force systems

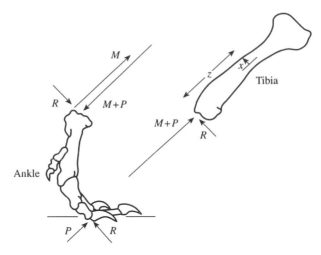

Fig. 19. Forces on foot and lower tibia of *Tyrannosaurus*

where x is the distance of the fibre from the centre line, measured in the same direction as R.

Sauropods, like *Apatosaurus*, and elephants each have long femurs and short toes, and may have moved their legs in the same way. Peak stresses at the front and back of the humerus of a running elephant have been estimated at $69 = -8 + 77$ and $-85 = -8 - 77$ newtons per square millimetre, indicating a dominance of bending (± 77) over compression (-8). It follows that, in a given fibre (given x), the key quantity in (15) is Rz/Z. Reasoning of this nature, together with the supposition that R is proportional to the weight w, led Alexander to define Z/wz as an *indicator of athletic ability*. The following values have been calculated for *femurs*, in units of square metres per giganewton, where 1 giganewton $= 10^9$ newtons, and with body mass given in tonnes, where 1 tonne $= 10^3$ kilograms.

	Body mass	Femur Z/wz
Ostrich	0.04	44
Buffalo	0.5	22
Triceratops	6–9	15–21
Human	0.06	15
Tyrannosaurus	8	9
Apatosaurus	34	9
Elephant	2.5	7
Diplodocus	12–19	3–5

These results indicate *quantitatively* how much less athletic are elephants than buffaloes, for example. Elephants neither gallop nor jump, which also describes their lack of athleticism in another way. *Diplodocus* was less athletic than either

Apatosaurus or the elephant, and probably could not even run. *Triceratops* seems as athletic as a buffalo on this measure, and perhaps could have galloped, as can a rhinoceros.

Solutions to Worksheet 1

1.

Diplodocus	18560 kg	*Triceratops*	6080 kg
Tyrannosaurus	116 cc	*Stegosaurus*	48 cc.

$$m = 91b \text{ because } 45 \times 45 \times 45 = 91\,125.$$

$$\frac{(91b - 64b) \times 100}{64b} = 42\%.$$

2.

c	-4	-3	-2	-1	0	1	2	3	4
$m = 0.25c^2$	4	2.25	1	0.25	0	0.25	1	2.25	4
$m = 0.5c^2$	8	4.5	2	0.5	0	0.5	2	4.5	8
$m = 0.1c^2$	1.6	0.9	0.4	0.1	0	0.1	0.4	0.9	1.6

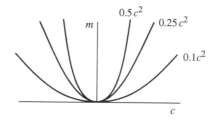

3. The circle has diameter $70/\pi = 22.28$ cm.

4.

b	2	2.73	3
Diplodocus	107	18163	121203
Triceratops	47	5932	35437
Stegosaurus	28	2941	16389

Very poor approximations are evidently obtained by using $b = 2$ or $b = 3$ in place of $b = 2.73$.

5. *Tyrannosaurus* 7328 kg   *Iguanodon* 5341 kg

6. For $b = 1$, $m = ac$ is satisfied by $c = 0$, $m = 0$.

The gradient between the origin and a typical point is therefore

$$\frac{m-0}{c-0} = a.$$

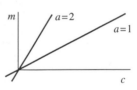

This is a constant for all points (c, m) and therefore we have a straight line.

(B) is $y = bx + \log a$, so that $y = \log a$ when $x = 0$.

Therefore the gradient

$$\frac{y - \log a}{x - 0} = b$$

is constant.

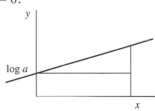

Solutions to Worksheet 2

1. Let r be the radius of the circle. Then $C = 2\pi r$ and $\frac{1}{2}A = \pi r^2$, so by eliminating r, $\frac{1}{2}A = \pi(C/2\pi)^2 = C^2/4\pi$.

   $s = w/A = 2\pi w/C^2$.

   $w = 10m$ and $20\pi = 63$ approximately, so $s = 63m/C^2$.

   Iguanodon has $s = \dfrac{63 \times 5341}{(0.57)^2} = 1\,035\,651$ newtons per square metre.

2. A force of $w/4$ acts through each leg for quadrupeds (as distinct from $w/2$ for bipeds in question 1), so $s = \frac{1}{4}w/\frac{1}{4}A = w/A$ with $\frac{1}{4}A = C^2/4\pi$ gives

   $$s = \pi w/C^2.$$

   When *metres* are consistently used for the values of C and for $g = 10$ metres per second per second, the following values of s are obtained in newtons per square metre.

Brachiosaurus	2 987 719	*Triceratops*	1 399 833
Diplodocus	1 788 842	*Stegosaurus*	1 158 019

3. $u = \dfrac{15}{(3000)^{1/2}} = 0.27$

   $v = 0.27(50)^{1/2} = 1.91$ metres per second

4. 1 metre per second $= \dfrac{60 \times 60}{1000}$ kilometres per hour

   $= \frac{5}{8} \times 3.6$ miles per hour $= 2.25\,\text{mph}$

5. ostrich $1.72 < r < 2.13$ human $3.23 < r < 4.61$
   dogs $1.44 < r < 7.65$ sheep $3.92 < r < 5.72$.

6. Using $g = 10$ metres per second per second

   $$v = \frac{(4 \times 0.4 \times 10)^{1/2}}{1.38}\left(\frac{180}{160} - 0.75\right) = 1.1 \text{ metres per second}.$$

7. My own dimensions with these speeds give the two lines

$$a = 2.32\,(1 - b)$$

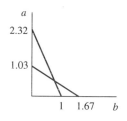

and $a = 0.62\,(1.67 - b)$ as shown. They intersect where $a = 0.57$, $b = 0.76$. The line $r = 0.57u + 0.76$ would be a fit of the human data in (u, r) space upon which my own data happen to lie exactly.

Acknowledgements

This masterclass is essentially based on the first four chapters of the book by the zoologist R. McNeill Alexander referenced below. Figures 13, 14 and 19 are adapted from his book. I am glad to acknowledge my indebtedness to the balanced account which it gives. As an applied mathematician, I have put my own gloss on certain aspects of the material, and I have devised the Worksheets.

I am grateful to Mr. Simon Johnson, of the Reading University Photographic Service, for being with me in the right place at the right time to take the photographs at my behest. Figures 1, 4 and 7 have appeared in the *University of Reading Bulletin and Annual Report*. The beam balance used in the weighing of Fig. 12 was kindly given to me by Mr. Stephen Samuels of Leighton Park School Physics Department.

References and Further Reading

R. McNeill Alexander, *Dynamics of dinosaurs and other extinct giants* (Columbia University Press, New York, 1989)

J. F. Anderson, A. Hall-Martin, and D. A. Russell, 'Long bone circumference in mammals, birds, and dinosaurs', *Journal of Zoology* (A) **207** (1985), 53–61

R. T. Bird, 'Did brontosaurus ever walk on land?', *Natural History*, New York **53** (1944), 60–7

R. Coria and L. Salgado, 'A new giant carnivorous dinosaur from the Cretaceous of Patagonia', *Nature* **377** (21 September, 1995) 224–226

L. B. Halstead and Jenny Halstead, *Dinosaurs* (Blandford Press, Poole, 1981)

M. Lockley, *Tracking dinosaurs* (Cambridge University Press, 1991)

R. A. Thulborn and M. Wade, 'Dinosaur trackways in the Winton formation (mid-Cretaceous) of Queensland', **21** (1985), 413–517

10 Pythagorean Number Triples

by DAVID STIRLING
Department of Mathematics, University of Reading

1. Introduction

The subject of this class is called number theory, which deals with the *whole numbers*, that is 1, 2, 3, 4 and so on and their negatives −1, −2, −3, ... We shall not be interested in fractions or decimal numbers. This is a very old and beautiful subject but it is rather different from the way we do ordinary calculations for we cannot always solve simple equations. For example, the equation $2x = 3$ does not have a solution if we only allow x to be a whole number.

Some positive whole numbers can be expressed as the product of two smaller positive whole numbers (for example $6 = 2 \times 3$), while others cannot (for example, 7). The numbers which *cannot* be factorized into the product of two smaller numbers are called *prime* numbers. The first few prime numbers are 2, 3, 5, 7, 11, 13, 17, ... A great deal is known about the prime numbers but a lot has yet to be discovered. For example, we know that the list of prime numbers is infinite, that is, it never ends. As we shall see, this is not difficult to prove. On the other hand, although the proportion of the numbers which are prime becomes smaller as the numbers become larger (that is there are more primes between 100 and 1000 than there are between 1 000 100 and 1 001 000), prime numbers do occur close together, two apart. (For, except for 2, all prime numbers must be odd). Thus 11 and 13, 17 and 19, 29 and 31, ... are so-called 'prime pairs' and so is 1019 and 1021. These prime pairs become less frequent as the numbers increase, but nobody knows if the list of them carries on for ever or if it stops after some point.

Division with Remainder

If a is a whole number greater than 1 we can divide any whole number x by a and obtain a *quotient q* and *remainder r*, so that

$$x = a \times q + r$$

where q and r are whole numbers and r is one of the *remainders* $0, 1, 2, \ldots, a - 1$. Thus r is at least 0 and less than a.

So if we divide 61 by 5 in this way we have

$$61 = 5 \times 12 + 1$$

giving a quotient of 12 and remainder 1 when we divide by 5.

We say that *a divides x* (or *is a factor of x*) if the remainder is zero when we apply this process of dividing x by a. Putting this another way, a divides x exactly when there is a whole number q with

$$x = a \times q.$$

Now a prime number cannot be factorized into the product of two smaller numbers, so if p is a prime number its only factors are itself and 1. If a number is not a prime number then it can be factorized, and each of the factors may be a prime number (in which case we cannot split it up) or it may be possible to factorize it. If we start with a number x and factorize it, then factorize the factors and so on, we will eventually have to stop once all the factors are prime numbers. This gives us:

Theorem 1

Every whole number greater than 1 is either a prime number or it is the product of several prime numbers, its prime factors. Apart from the order in which we write them, the prime factors are unique.

Thus $60 = 2 \times 30$.

2 is a prime number but $30 = 2 \times 15$, and $15 = 3 \times 5$.

So $60 = 2 \times 2 \times 3 \times 5$ and we have gone as far as we can, since 2, 3 and 5 are prime numbers.

Given Theorem 1 we can prove that the list of prime numbers never ends.

Theorem 2

There are infinitely many prime numbers, that is, no matter how many we have found there are more.

Proof

Suppose we have found n prime numbers, which we shall call $p_1, p_2, p_3, \ldots, p_n$. Then consider the number

$$N = (p_1 \times p_2 \times p_3 \times \cdots \times p_n) + 1.$$

This is bigger than 1 so it has a prime factor, p say. (p could be N itself.) Now when we divide N by p_1, we have a remainder of 1 so p_1 is not a factor of N. For the same reason p_2 is not a factor of N, nor are $p_3, p_4, \ldots, p_n$. That is, p is a prime factor of N that cannot be equal to any prime number we have found already, so it is another prime number.

That is, no matter how many prime numbers we have found there are more to be found! □

Common Factors

Given two positive whole numbers they may have a common factor, that is, a number which divides both of them: although 4 does not divide 6 nor does 6 divide 4, they have a common factor of 2. If we accept 1 as a factor (a rather boring one) then every pair of numbers has a common factor, maybe many of them. The most interesting common factors are the larger ones, in particular the largest. We shall denote the *highest common factor* (HCF) of a and b by $\text{HCF}(a, b)$.

Example

Let $a = 42$ and $b = 60$. Then (by testing the numbers in turn) the common factors of a and b are 1, 2, 3 and 6 so the HCF is 6.

An interesting feature about the highest common factor of x and y is that it can be expressed in terms of whole–number multiples of x and y.

Theorem 3

If d is the highest common factor of a and b then there are whole numbers x and y for which

$$d = ax + by.$$

(The proof is technical; for those interested it is in the Appendix.)

Notice that the HCF of 42 and 60 is 6 (above) and

$$6 = 42 \times 3 + 60 \times (-2).$$

This leads us to a way of finding highest common factors, called Euclid's Algorithm. (Euclid was one of the ancient Greek mathematicians, whose books on geometry were very famous for hundreds of years. He lived about 300BC.)

Euclid's Algorithm

To find the HCF of two positive whole numbers a_1 and a_2, where we choose a_1 to be the larger of the two if they are not equal. Divide a_1 by a_2 to obtain the quotient q_2 and remainder a_3, so

$$a_1 = a_2 q_2 + a_3$$

where a_3 is a whole number, $0 \leq a_3$ and $a_3 < a_2$.

Now divide a_2 by a_3, with quotient q_3 remainder a_4 so that

$$a_2 = a_3 q_3 + a_4, \quad 0 \leq a_4, a_4 < a_3.$$

Continue this way. The numbers $a_1, a_2, a_3, \ldots$ are all whole numbers and they become smaller at each step so eventually we must get to zero. That is, at some step (the nth say), $a_n = 0$. (This could occur with $n = 3$.)
    Then

$$a_{n-2} = a_{n-1} q_{n-1} + 0.$$

The HCF is a_{n-1}.

Example

Find the HCF of 294 and 798.

Let $a_1 = 798$ and $a_2 = 294$.

Then $798 = 294 \times 2 + 210.$ $\qquad (a_3 = 210)$

$294 = 210 \times 1 + 84$
$(a_2 = a_3 \times q_3 + 84)$ $\qquad (a_4 = 84)$

$210 = 84 \times 2 + 42$ $\qquad (a_5 = 42)$

$84 = 42 \times 2 + 0$ $\qquad (a_6 = 0)$

Therefore HCF $= 42$.

Check: $\qquad\qquad\qquad 798 = 42 \times 19$ and $294 = 42 \times 7$.

So 42 is a common factor. By working out the factors of 294 we see that these are $1, 2, 3, 6, 7, 14, 21, 42, 49, 98, 147$ and the last three are not factors of 798.

Worksheet 1

1. Find the HCF of the following pairs of numbers using Euclid's Algorithm:

   4 and 6; $\qquad$ 13 and 17; $\qquad$ 8 and 20;
   72 and 128; $\qquad$ 72 and 96.

2. Factorize the numbers concerned into their *prime* factors and use these to find the highest common factors of the pairs in question 1. (Make sure your answers agree!)

   (The numbers 2, 3, 5, 7, 11, 13, 17 and 19 are prime numbers.)

3. A *multiple* of a number x is a number of the form xy where y is a whole number. A *common multiple* of the numbers a and b is a number which is both a multiple of a and of b. Thus 54 is a common multiple of 6 and 9, and so is any multiple of 54. There are also smaller common multiples of 6 and 9. Find the lowest common multiple of all the pairs in question 1.

4. Do you see any connection between the highest common factor of two numbers and their lowest common multiple? Can you convince a friend that this connection is always true?

2. Diophantine equations

Let a and b be two positive whole numbers and d be their HCF. Then we know that d is of the form

$$d = ax + by$$

for some whole numbers x and y. Now suppose that c is another common factor of a and b (not necessarily the highest). Then $a = ca_1$ and $b = cb_1$ for some whole numbers a_1 and b_1. From this

$$d = ax + by = ca_1 x + cb_1 y = c(a_1 x + b_1 y)$$

so that d is a multiple of c.

We have proved:

Theorem 4

Every common factor of two numbers divides exactly into the highest common factor.

I now want to consider 'Diophantine Equations'. This impressive name just means that we want to consider equations where the solutions are required to be whole numbers. They are named after Diophantus, another ancient Greek mathematician. We have already seen that some equations (such as $2x = 3$) do not have whole number solutions so solving diophantine equations is not as simple as we might wish. The equation $ax = b$ is not, however, too hard to solve, for if it has a solution x which is a whole number, then b is a multiple of a. Also if b is a multiple of a then there is a whole-number solution x.

Notice the logic here. We have said that the equation $ax = b$ has a whole-number solution if b is a multiple of a. We also said that if $ax = b$ has a whole-number solution then b is a multiple of a, or putting the same thing another way, $ax = b$ has a whole-number solution only if b is a multiple of a. That is, the two properties (i) that $ax = b$ has a whole-number solution and (ii) that b is a multiple of a are true for exactly the same numbers b. In other words they are both true, or neither of them is true. This is often said mathematically as '$ax = b$ has a whole-number solution if and only if b is a multiple of a'.

Let us see what we already know about diophantine equations. Theorem 3 told us that if d is the HCF of a and b then there are whole numbers x_1 and y_1 with $d = ax_1 + by_1$. So we know that if $c = d$ then the diophantine equation

$$ax + by = c$$

has a solution.

In fact we know more than this, for Euclid's Algorithm allows us to find the solution.

Example

Let $a = 72$ and $b = 128$. Then if we use Euclid's Algorithm to find the HCF we have

$$128 = 72 \times 1 + 56 \tag{1}$$

$$72 = 56 \times 1 + 16 \tag{2}$$

$$56 = 16 \times 3 + 8 \tag{3}$$

$$16 = 8 \times 2 + 0 \tag{4}$$

Therefore HCF = 8.

Now $8 = 56 - 16 \times 3$ (equation (3))

and equation (2) shows us that $16 = 72 - 56 \times 1$, so

$$8 = 56 - (72 - 56 \times 1) \times 3 = 56 \times 4 + 72 \times (-3).$$

So
$$8 = 56 \times 4 + 72 \times (-3)$$
$$= (128 + 72 \times (-1)) \times 4 + 72 \times (-3)$$
$$= 128 \times 4 + 72 \times (-4 - 3)$$
$$= 128 \times 4 + 72 \times (-7).$$

Therefore

$$128x + 72y = 8$$

has solution $x = 4$, $y = -7$. It may have other solutions, too.

Now let us look at the equation

$$ax + by = c \qquad (*)$$

more carefully.

We know that there are whole numbers x_1 and y_1 for which $ax_1 + by_1 = d$, where $d = \text{HCF}(a, b)$. Then we can see easily that if c is a multiple of d, equation ($*$) has a solution. (For if $c = de$, then $x = x_1 e$, and $y = y_1 e$ satisfies equation ($*$).) So we know some circumstances when there is a solution.

Now suppose that c is a number for which equation ($*$) has a solution. Let d be the HCF of a and b. Then $a = da_1$ and $b = db_1$ for some whole numbers a_1 and b_1. Then if x and y satisfy equation ($*$)

$$c = ax + by = da_1 x + db_1 y$$
$$= d(a_1 x + b_1 y).$$

Because $a_1 x + b_1 y$ is a whole number, we deduce that c is a multiple of d.

We have proved that if c is a multiple of d then equation ($*$) has a solution and that if equation ($*$) has a solution then c is a multiple of d. That is, we have proved the following result.

Theorem 5

The diophantine equation

$$ax + by = c$$

has a solution if and only if c is a multiple of the HCF of a and b.

This equation, if it has solutions at all, will have many solutions.

Example

$$20x + 8y = c$$

Since $\text{HCF}(20, 8) = 4$ the equation has a solution if and only if c is a multiple of 4.

Now find a solution (by Euclid's Algorithm or by guessing it) of

$$20x + 8y = 8.$$

Since $4 = 20 - 8 \times 2$, $\qquad 8 = 20 \times 2 + 8 \times (-4)$

so $x = 2$, $y = -4$ is a solution.

However

$$20 \times (-8) + 8 \times (20) = 0$$

and for all z

$$20 \times (-8z) + 8 \times (20z) = 0 \times z = 0,$$

so that $20 \times (2 - 8z) + 8 \times (-4 + 20z) = 8 - 160z + 160z = 8.$

Therefore other solutions of $20x + 8y = 8$

are
$$
\begin{array}{llr}
x = -6, & y = 16 & (z = 1) \\
x = 10, & y = -24 & (z = -1) \\
x = -14, & y = 36 & (z = 2)
\end{array}
$$

and so on, for all whole number values of z.

Other diophantine equations are equally interesting, and some are fascinatingly complicated. For example, if we consider

$$x^2 - y^2 = a$$

(where x^2 means $x \times x$) then it is not hard to find a solution if $a = 1$. $a = 2$ is more awkward but $a = 3$ isn't too difficult. In fact there is no solution if $a = 2$, for then

$$x^2 = y^2 + 2.$$

Let $x = y + z$ so that
$$
\begin{aligned}
x^2 &= (y + z) \times (y + z) \\
&= y^2 + 2yz + z^2.
\end{aligned}
$$

Then $z^2 + 2yz = 2$ from which we see that

$$z^2 = 2 - 2yz = \text{an even number}.$$

Now the square of an odd number is odd, so z must be even (for it cannot be odd), so z^2 is a multiple of 4. But if z is even, $2yz$ is also a multiple of 4 and so

$$2 = z^2 + 2yz = \text{a multiple of 4}.$$

This is nonsense, for 2 isn't a multiple of 4. So if we assume the equation has a solution we deduce some nonsense—so there cannot be a solution.

The equation I want to consider in detail, however, is slightly different:

$$x^2 + y^2 = z^2.$$

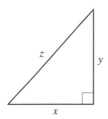

Fig. 1. Pythagoras's Theorem

The interest in this is from *Pythagoras's Theorem*, that if x, y and z are the sides of a right-angled triangle as shown, then $x^2 + y^2 = z^2$. I want to find all the whole-number solutions.

Notice that you could use this to draw a right angle: $3^2 + 4^2 = 5^2$ so if you draw a triangle with $x = 3$, $y = 4$ and $z = 5$, then it has a right angle.

Worksheet 2

1. Find a solution, with x and y whole numbers, of
$$3x + 4y = 1.$$

2. Find a whole-number solution of
$$2x + 4y = 6.$$

3. Show that the equation
$$2x + 4y = 3$$
does not have any solutions where x and y are whole numbers.

4. Suppose that u and v are whole numbers which satisfy
$$3u + 4v = 0.$$

Show that v must be a multiple of 3; call it $3z$.

Use this to show that for some whole number z
$$u = -4z \text{ and } v = 3z.$$

5. Look at your solution of question 1, and choose two whole numbers a and b with $3a + 4b = 1$.

Now suppose that x and y also satisfy
$$3x + 4y = 1.$$

Show that
$$3(x - a) + 4(y - b) = 0$$
and use question 4 to show that, for some whole number z
$$x = a - 4z$$
$$y = b + 3z.$$

6. Suppose we want to find whole number solutions of

$$2x^2 + 5xy + 2y^2 = 2.$$

We can factorize the left-hand side to get

$$(2x + y)(x + 2y) = 2.$$

Now we find a whole-number solution of the simultaneous equations

$$2x + y = 2$$
$$x + 2y = 1.$$

Check that your answer satisfies $2x^2 + 5xy + 2y^2 = 2$.

If, instead, we set

$$2x + y = 1$$
$$x + 2y = 2$$

(so that $(2x + y)(x + 2y) = 2$ again) what solutions do you find?

7. Try the same thing on

$$(2x + y)(x + 2y) = 8,$$

noticing that there are several ways in which 8 can be written as the product of two whole numbers ab, and solve

$$2x + y = a$$
$$x + 2y = b.$$

8. Now look at

$$(2x + y)(x + 2y) = 9.$$

9. (Hard!) Suppose we want to find all the whole-number solutions of

$$x^2 - y^2 = a. \tag{1}$$

We have already seen that this has a solution if $a = 1$ or $a = 3$ but not if $a = 2$. Write $a = bc$ (where, as always, b and c are whole numbers) and consider the equations

$$x + y = b \tag{2}$$
$$x - y = c. \tag{3}$$

Show that if these two equations have whole-numbers solutions, x and y, then these values of x and y satisfy $x^2 - y^2 = a$.

By finding the solutions of equations (2) and (3), show that there are whole number solutions if b and c are both odd numbers, or if b and c are both even. Show also that equations (2) and (3) do *not* have whole-number solutions if one of b and c is odd and the other even.

Use this to show that if a is odd, then both its factors b and c (including the case $b = a$, $c = 1$) are odd and therefore equation (1) has whole-number solutions.

Show also that if a is a multiple of 4, then it can be factorized into b and c where both b and c are even.

We have shown that equation (1) has whole-number solutions if a is either odd or is a multiple of 4.

Try a few examples (e.g. $a = 4$, 5 or 48) to see how many solutions there are.

3   Pythagorean triples

Let us devote our attention to the equation

$$x^2 + y^2 = z^2$$

where we want x, y and z to be positive whole numbers. Now trial and error will give us one solution

$$x = 3, \quad y = 4, \quad z = 5.$$

It is not hard to see how to make new solutions from this, particularly if we remember the geometry. If Fig. 2

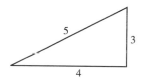

Fig. 2.   Pythagorean Triangle

gives us a right-angled triangle then the triangle whose sides are twice those shown will also be right-angled. We then check that $6^2 + 8^2 = 10^2$.

More generally

$$(ax)^2 + (ay)^2 = a^2(x^2 + y^2) = (az)^2.$$

This is not giving us very different solutions: they are the sides of triangles of the same 'shape'. (Technically we call them similar triangles.)

Instead of multiplying by a we could divide by a, if the numbers x, y and z have a common factor. This would return us to the basic 3, 4, 5 solution (for 3, 4 and 5 have no common factors). So let us look for new answers and ask:

Find all the solutions of

$$x^2 + y^2 = z^2$$

where x, y and z are positive whole numbers where x and y have no common factor greater than 1.

Now if $x^2 + y^2 = z^2$ and x and y have no common factor they cannot both be even. Less obviously, they cannot both be odd either, for if they were we would have

$$x = 2s + 1, \qquad y = 2t + 1$$

for some whole numbers s and t. Then

$$x^2 + y^2 = 4s^2 + 4s + 1 + 4t^2 + 4t + 1$$
$$= 4(s^2 + s + t^2 + t) + 2$$

so that $z^2(-x^2 + y^2)$ is even, but not a multiple of 4 for its remainder when we divide by 4 is 2.

But the square of an odd number is odd, so z cannot be odd, and the square of an even number is a multiple of 4 (for $(2r)^2 = 4r^2$) so z cannot be even either. Therefore

one of x and y is odd and the other is even.

For simplicity suppose that x is odd and y is even. Then if $x^2 + y^2 = z^2$, z^2 must be odd, so z is odd.

Now $y^2 = z^2 - x^2 = (z + x)(z - x)$ and because x and z are both odd, $z + x$ and $z - x$ are both even, and

$$\tfrac{1}{4}y^2 = \left(\frac{z + x}{2}\right)\left(\frac{z - x}{2}\right) = ab, \quad \text{say}$$

where $a = \tfrac{1}{2}(z + x)$ and $b = \tfrac{1}{2}(z - x)$.

At this point we need the tricky step! We know that ab is a square $(\tfrac{1}{2}y)^2$ but we do not know whether a and b are themselves squares. (For $2 \times 18 = 36$ is a square but 2 and 18 are not.) To show that a and b are squares we need to know that they have no common factor greater than 1. Why?

Let $\tfrac{1}{2}y = p_1 p_2 \ldots p_n$ be its factorization into *prime* numbers, so that

$$ab = \left(\tfrac{1}{2}y^2\right) = p_1^2 p_2^2 \ldots p_n^2.$$

Therefore in the product ab the prime number p_1 occurs twice. If it occurred once in a and once in b then p_1 would be a common factor of a and b. Putting this another way, if a and b have no common factor greater than 1 then both the p_1s occur in a or they both occur in b. The same applies to p_2, p_3 and so on. That is, a is the product of the squares of some of the prime numbers, and b is the product of the rest of the prime numbers squared. This would show that a and b are squares. Therefore

if a and b have no common factor greater than 1
then a and b are squares.

We have to show that a and b have no common factor other than 1. Suppose that p is a prime number which divides a and b. Then p divides $a + b$ and $a - b$. That is, p divides z and x, for $a + b = \tfrac{1}{2}(z + x) + \tfrac{1}{2}(z - x) = z$, and $a - b = x$. Therefore p^2 divides z^2 and x^2.

Now $y^2 = z^2 - x^2$ so that p^2 divides y^2 and so p divides into y. That is, p is a common factor of x and y which we disallowed at the beginning.

To sum up: if x and y have no common factor greater than 1, then they have no prime common factor. (Any common factor greater than 1 must be divisible by a prime.) Then

$$\left(\tfrac{1}{2}y\right)^2 = ab$$

where a and b are squares. Call them s^2 and t^2 where s and t are positive. Since a and b have no common factor, neither do s and t.

Now
$$x = a - b = s^2 - t^2$$
$$y^2 = 4ab = 4s^2t^2 \qquad \text{so } y = 2st \text{ (if } y \text{ is positive)}$$
$$z = a + b = s^2 + t^2.$$

That is, if x, y and z are solutions of

$$x^2 + y^2 = z^2$$

where x and y have no common factor greater than 1, then for some whole numbers s and t, with no common factor greater than 1

$$x = s^2 - t^2,$$
$$y = 2st,$$
$$z = s^2 + t^2.$$

Since z is odd one of s and t is odd and the other is even; also s is greater than t since x is to be positive.

That is the hard work done! We can easily check that the x, y, z we have found always do satisfy the equation for

$$x^2 + y^2 = \left(s^2 - t^2\right)^2 + \left(2st\right)^2$$
$$= s^4 - 2s^2t^2 + t^4 + 4s^2t^2$$
$$= s^4 + 4s^2t^2 + t^4$$
$$= \left(s^2 + t^2\right)^2$$
$$= z^2.$$

The first few Pythagorean triples are (using the fact that s and t have no common factor greater than 1 and one is odd, the other even.)

s	t	x	y	z
2	1	3	4	5
3	2	5	12	13
4	1	15	8	17
4	3	7	24	25
5	2	21	20	29
5	4	9	40	41
6	1	35	12	37
6	5	11	60	61

Under the conditions we have given there are no repetitions.

Notice what we have done. By going carefully through all the logic the way we have done, we have shown that every solution of the equation $x^2 + y^2 = z^2$ where x, y and z are whole numbers with no common factor is of the form we have given (and that these are solutions). We have found them all! If we remove the restriction that x and y have no common factor, then all other solutions are obtained from the ones we have found by multiplying x, y and z throughout by the same whole number.

The equation

$$x^3 + y^3 = z^3$$

is very different, however. It has no solutions at all if x, y and z are positive whole numbers.

The 17th-century mathematician Pierre de Fermat wrote in the margin of a book that if n is at least 3

$$x^n + y^n = z^n$$

has no positive whole-number solutions. This is a very famous problem, 'Fermat's Last Theorem', which has finally been solved; a British mathematician, Andrew Wiles, announced a proof in the summer of 1993. It only took three centuries!

Solutions to Worksheet 1

1.

Numbers	HCF
4 and 6	2
13 and 17	1
8 and 20	4
72 and 128	8
72 and 96	24

2. $4 = 2 \times 2$, $\quad 6 = 2 \times 3$: they have one 2 in common.

$13 = 13$ (prime), $\quad 17 = 17$ (prime): no common factor bigger than 1.

$8 = 2 \times 2 \times 2$, $\quad 20 = 2 \times 2 \times 5$: two 2s in common.

$72 = 2 \times 2 \times 2 \times 3 \times 3$, $\quad 128 = 2 \times 2 \times 2 \times 2 \times 2 \times 2 \times 2$: three 2s in common.

$72 = 2 \times 2 \times 2 \times 3 \times 3$, $\quad 96 = 2 \times 2 \times 2 \times 2 \times 2 \times 3$: three 2s and a 3 in common.

3.

Numbers	Lowest Common Multiple
4 and 6	12
13 and 17	221
8 and 20	40
72 and 128	1152
72 and 96	228

4. If the numbers are a and b, then

$$\text{HCF} \times \text{LCM} = a \times b$$

(where LCM means lowest common multiple).

Why? Suppose a_1 and b_1 have HCF 1, that is, no number larger than 1 is a common factor. Then the LCM of a_1 and b_1 is $a_1 \times b_1$. (For any smaller multiple of b_1 than $a_1 \times b_1$ must omit some of the prime factors of a_1 and not be a multiple of a_1, hence not a common multiple of a_1 and b_1.) Now let $\mathrm{HCF}(a, b) = d$. Then $a = da_1$ and $b = db_1$ where $\mathrm{HCF}(a_1, b_1) = 1$. Then $ab = d^2 a_1 b_1$ and $\mathrm{LCM}(a, b) = \mathrm{LCM}(da_1, db_1) = d \times \mathrm{LCM}(a_1, b_1) = da_1 b_1$, so

$$ab = d \times \mathrm{LCM}(a, b) = \mathrm{HCF} \times \mathrm{LCM}.$$

Solutions to Worksheet 2

1. $x = -1$, $y = 1$ will do, but you may find others.
2. $x = 1$, $y = 1$ is a solution. Again, you may find others.
3. If x and y are whole numbers then $2x$ and $4y$ are both multiples of 2, so that $2x + 4y$ is a multiple of 2.
   Because 3 is *not* a multiple of 2, $2x + 4y$ cannot equal 3 if x and y are whole numbers.
4. $3u + 4v = 0$ shows that $4v = -3u$ so that $4v$ is a multiple of 3. If we write it out in its prime factors, then, 3 will be one of them. But that factor cannot come from the 4 (for its prime factors are 2 and 2) so it comes from the v, that is, v has to be a multiple of 3.

   Then $v = 3z$ for some whole number z, so

   $$-3u = 4v = 12z$$

   giving $u = -4z$.
5. From our question 1, $a = -1$ and $b = 1$ (but choose your own a and b if you found a different answer to question 1).

   Then suppose that $\qquad 3x + 4y = 1$ and we have

   $$3(-1) + 4(1) = 1.$$

   Subtracting the two equations gives

   $$3x - 3(-1) + 4y - 4(1) = 1 - 1 = 0$$

   so

   $$3(x + 1) + 4(y - 1) = 0.$$

   Using question 4, then, with $u = x + 1$ and $v = y - 1$ we have some whole number z for which $u = x + 1 = -4z$ and $v = y - 1 = 3z$ so

   $$x = -1 - 4z$$
   $$\text{and} \quad y = 1 + 3z.$$

   (If you used different numbers a and b from that, you can check that your a and b are of the form

   $$a = -1 - 4z$$
   $$b = 1 + 3z$$

   for some whole number z.)

6. If $2x + y = 2$

   and $x + 2y = 1$

   then $y = 2 - 2x$ from the first equation so

   $$x + 2(2 - 2x) = 1$$

   or $-3x = -3.$

   This gives $x = 1$. The first equation now gives $y = 0$.

   We have the solution $x = 1$, $y = 0$.

   If instead you set $2x + y = 1$

                               $x + 2y = 2$

   you get $x = 0$, $y = 1$.

7. We can write $8 = ab$ with $a = 8$, $b = 1$ or $a = 4$, $b = 2$, or $a = 2$, $b = 4$ or $a = 1$, $b = 8$.

   Try the first of these:

   $$2x + y = 8$$
   $$x + 2y = 1.$$

   Then $y = 8 - 2x$ so (putting this in the second equation)

   $$x + 2(8 - 2x) = 1$$

   or $-3x = -15.$

   This gives $x = 5$ and $y = -2$ (from the first equation).

   Now try $2x + y = 4$

                              $x + 2y = 2$

   so that $y = 4 - 2x$ and $x + 2(4 - 2x) = 2$ giving

   $$-3x = -6.$$

   In this case we get $x = 2$ and from the first equation $y = 0$.

   $$a = 2, b = 4 \text{ gives } x = 0, \quad y = 2.$$
   $$a = 1, b = 8 \text{ gives } x = -2, \quad y = 5.$$

   So the solutions we have found are:

x	5	2	0	-2
y	-2	0	2	5

8. We can write $9 = ab$ with $a = 9$, $b = 1$, or $a = 3$, $b = 3$ or $a = 1$, $b = 9$.

   Now $2x + y = 9$

   and $x + 2y = 1$

gives $y = 9 - 2x$, so putting this into the second equation,

$$x + 2(9 - 2x) = 1$$
$$\text{or} \qquad -3x = -17$$

which has no whole-number solution, for 17 isn't a multiple of 3.

If $$2x + y = 3$$
$$\text{and} \qquad x + 2y = 3$$

then $y = 3 - 2x$ so from the second equation

$$x + 2(3 - 2x) = 3$$

giving $$-3x = -3.$$

Therefore $x = 1$ and, from the first equation $y = 1$.

If $$2x + y = 1$$
and $$x + 2y = 9$$

then $y = 1 - 2x$ so $x + 2(1 - 2x) = 9$ giving $-3x = 7$ which has no solution. The only solution is $x = 1$, $y = 1$.

9. If we have whole-number solutions of

$$x + y = b$$
$$x - y = c$$

where $a = bc$, then

$$x^2 - y^2 = (x + y)(x - y) = bc = a,$$

so x and y satisfy equation (1).

If x and y satisfy equations (2) and (3), then adding and subtracting the two equations gives

$$2x = b + c$$
$$2y = b - c$$

so that $b + c$ and $b - c$ must be multiples of 2. In this case $x = \frac{1}{2}(b + c)$ and $y = \frac{1}{2}(b - c)$, where x and y are whole numbers and they satisfy equations (2) and (3).

Therefore there are solutions if b and c are either both even or both odd. Also, if one of b and c is even and the other odd then $b + c$ is odd, so there is no whole number x with $2x = b + c$, so equations (2) and (3) do *not* have a solution in this case.

Let a be an odd number. Then we can factorize a into $a \times 1$ ($b = a$, $c = 1$) and both b and c are odd so that equation (1) will have at least one solution. (It may have several, for there may be different ways of factorizing a, always into two odd numbers).

If a is a multiple of 4, then we can write $a = 4d$ where d is a whole number. Factorize d in any way you like so that $d = ef$ where e and f are whole numbers (e.g. $e = d$, $f = 1$). Then putting $b = 2e$, $c = 2f$ gives $a = bc$ where b and c are both even. Again equation (1) will have whole-number solutions.

If $a = 4$ then we can write it as 2×2 ($b = 2$, $c = 2$) and this is the only way of choosing b and c so that $bc = 4$ and b and c are either both even or both odd. This gives $x = 2$, $y = 0$, the only solution of $x^2 - y^2 = 4$ (where x and y are non-negative whole numbers).

If $a = 5$ and $b = 5$, $c = 1$ gives $x = 3$ and $y = 2$. These are the only non-negative integer solutions. (Putting $b = 1$, $c = 5$ gives $x = 3$, $y = -2$.)

If $a = 48$ then we can choose $b = 24$, $c = 2$ or $b = 12$, $c = 4$ or $b = 8$, $c = 6$. (These are the only possibilities with b at least as large as c and $b + c$ even.) These give $x = 13$, $y = 11$ or $x = 8$, $y = 4$ or $x = 7$, $y = 1$ respectively.

Appendix

Proof of Theorem 3 (If a and b are positive whole numbers then there are whole numbers x and y for which $\mathrm{HCF}(a, b) = ax + by$.)

Let A be the set of all *positive* numbers of the form $ax + by$ where x and y are whole numbers. Then since $a = a \times 1 + b \times 0$ and $b = a \times 0 + b \times 1$, a and b are members of A. Let d_1 be the *smallest positive number in A*.

Then by definition

$$d_1 = ax_1 + by_1$$

for some whole numbers x_1 and y_1 (because d_1 is in A).

Now let z be a member of the set A, so that $z = ax + by$ for some whole numbers x and y. We divide z by d_1 so that

$$z = d_1 q + r$$

for some whole number q and remainder r with $r > 0$ and $r < d_1$ (division with the remainder).

Then
$$r = z - d_1 q$$
$$= (ax + by) - (ax_1 + by_1)q$$
$$= a(x - qx_1) + b(y - qy_1).$$

However, we know $r < d_1$ and d_1 was the smallest positive number in the set A, so r cannot be in the set A, for then it would be smaller than the smallest possible member of A. But $(x - qx_1)$ and $(y - qy_1)$ are whole numbers and $r = a(x - qx_1) + b(y - qy_1)$ so that if r is not in A it cannot be positive. Now we know that $r \geq 0$ and r is not positive, so $r = 0$. That is, z is a *multiple of d_1*.

We have shown that every member of A is a multiple of d_1.

Now let d be the HCF of a and b. So a is a multiple of d and so is b. Therefore d_1, which equals $ax_1 + by_1$, is a multiple of d. (For if $a = da_1$ and $b = db_1$, $d_1 = d(a_1x_1 + b_1y_1)$.) Therefore $d_1 \geq d$. (All positive multiples of d are greater than or equal to d.)

On the other hand, a and b are in the set A, as we saw, and d_1 divides every member of A. Therefore d_1 divides a and d_1 divides b, so d_1 is a common factor of a and b. But d is the *greatest* common factor, so $d_1 \leq d$.

We have shown that $d_1 \leq d$ and $d_1 \geq d$ so $d_1 = d$. Since $d_1 = ax_1 + by_1$ we have shown that

$$d = ax_1 + by_1$$

for some whole numbers x_1 and y_1.

Further Reading

A good starting point, for number theory as well as much of the rest of mathematics, is the most recent edition of the all-time classic:

R. Courant and H. Robbins, *What is mathematics* (Oxford University Press, 1941)

For more detail, look at:

H. Davenport, *The higher arithmetic*, 6th edition (Cambridge University Press, 1992)

An interesting and ancient account of sums of squares and related topics is:

Leonardo Pisano Fibonacci, *The book of squares*, annotated translation by L. E. Sigler (Academic Press, 1987)

For more advanced number theory, see

H. E. Rose, *A course in number theory*, 2nd edition (Oxford University Press, 1994)

11 Plato, Polyhedra and Weather Forecasting

by ANDY WHITE
Meteorological Office College, Reading

1. Setting the scene

Polyhedra are the main theme of this article; I'll explain what they are very soon. *Weather forecasting* motivates what we're going to do, and I'll say something about it in Section 2. As we shall see later, *Plato* is associated with an important particular sort of polyhedron. We shall also meet Archimedes, Euclid and Euler.

We have three aims.

1. To enjoy ourselves.
2. To learn some surprising and beautiful things about polyhedra.
3. To find that many scientific problems prompted by the world around us can be tackled either by careful mathematical reasoning or by careful visualization. If we're willing to use both methods, it's as if we each had the use of two brains, and our potential for solving mathematical/scientific problems goes up by a large factor (certainly much more than two).

What *is* a polyhedron?

A polyhedron is a three-dimensional figure

bounded by many ('poly') flat faces ('hedra').

(It's implicit in this definition that a polyhedron encloses a non-zero but finite volume. So 'many' means '4 or more'.)

Because each face of a polyhedron is flat, or *planar*, and planes intersect (if at all) in straight lines, each face of a polyhedron is a *polygon* of some sort. A polygon is a plane figure bounded by straight lines.

If you find these definitions a little confusing or strange (or both), don't worry. Soon you'll be in no doubt about what polyhedra are. If you were attending the class which this article is based on, I would hand you some polyhedra (made of clay, wax, wood, glass, cardboard or paper) and everything would be clear. Here we'll have to use two dimensional diagrams instead. The cube shown in Fig. 1 is a three-dimensional object (in real life) and it is *bounded* by its 6 square faces, which are flat. So it is a polyhedron. The pyramid, which has 4 triangular faces and one square one, is a polyhedron too. I think you will agree that the brick (which is a traditional builder's brick) is also a polyhedron; but the deep chunk

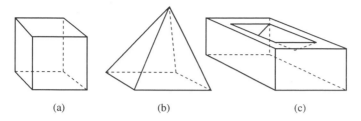

(a)   (b)   (c)

Fig. 1. (a) A cube, (b) a pyramid, and (c) a brick

out of its top surface puts it in a different class from that of the cube and the pyramid. The cube and the pyramid are each entirely on one side of every planar face, and are said to be *convex* polyhedra; but the brick is not entirely on one side of every face, and it is said to be a *concave* polyhedron. In practical terms, you can hold each face of a convex polyhedron flush against a (larger) flat table or wall, but you can't do this with a concave polyhedron – without deforming it or breaking it, that is. In fact, all the polyhedra we shall meet from now on will be convex ones, because they happen to be more relevant to our context – which I shall now tell you about.

2. Weather forecasting

When you see a weather forecast presented on TV, or hear a weather forecast on the radio or read one in a newspaper, it has not been arrived at entirely by intuition and applied human experience (although both of these processes have played a part).

What has happened is that a large computer model has been used to calculate the future state of the atmosphere from its known state some while ago. If the weather forecast was for today, and you were watching it being presented on UK television yesterday evening, the known state of the atmosphere from which the computer calculations were started was the one at noon (1200) yesterday. Thousands of measurements of temperatures, wind speeds, wind directions and many other quantities were taken in from around the world; they were used to define (as accurately as possible) the state of the atmosphere at noon yesterday.

One type of computer model represents the state of the atmosphere at thousands of *points* located over the globe of the Earth. At each point, quantities such as temperature, humidity, wind speed and wind direction are forecast. There are many layers of these points – typically about 20 – so there are altogether a very large number indeed. For example, in the UK Meteorological Office's global forecasting model there are 62 496 at each of 19 levels, making a total of 1 187 424 forecasting points.

The number of forecasting points is limited by computer speed and capacity, together with the obvious requirement that the numerical forecast should be produced long enough before the weather actually happens!

   The laws of classical physics (such as Newton's Second Law of Motion and the First Law of Thermodynamics) are bristling with information about *rates of change* of various quantities, so they are a very convenient basis for forecasting. From the state of the atmosphere at 1200 yesterday, the laws of physics were used by the computer model to calculate what the state of the atmosphere at 1210 yesterday is likely to have been. Then the calculation was repeated with the new (1210) values to find what the state of the atmosphere at 1220 yesterday is likely to have been. And so on, by ten-minute steps, until the likely state of the atmosphere at (say) 3 o'clock this afternoon had been calculated.

   The whole procedure is subject to error because:

(i) you never know the initial state of the atmosphere precisely;

(ii) you can't represent the state of the atmosphere completely even with meteorological quantities defined at 1 187 424 points!

To make the second aspect clear, recall that there are 62 496 forecasting points at each level of the Met. Office's global model. The Earth is approximately a sphere of radius (r) 6.37×10^6 metres, so its surface area $(4\pi r^2)$ is about 5.1×10^{14} square metres, and the mean area represented by each of the 62 496 points is equivalent to a square of side 9.03×10^4 metres, or about 90 kilometres.

   Now a lot of important things go on in the atmosphere on a much smaller horizontal *space scale* than 90 kilometres – just think of cumulus clouds or thunderclouds – so there is a major difficulty. Cumulus clouds and thunderclouds (for example) cannot themselves be represented in the computer model, but their *effects* are often important and they have to be allowed for if the model is to work reliably. The effects which have to be allowed for include the movement of heat, moisture and momentum which cumulus clouds and thunderclouds (for example) accomplish in the atmosphere. Representing these things well is a challenging problem.

   As well as its global model, the UK Met. Office has two regional models which are routinely used: the limited area model (LAM) and the mesoscale model. The LAM, in particular, is very important in day-to-day weather forecasting. These regional models have their forecasting points closer together than the global one does – down to a mean separation of about 17 kilometres in the mesoscale model. This goes some way towards removing the difficulty about small-scale, unresolved motions, but by no means removes it completely.

   The necessarily approximate nature of weather forecasting by computer means that the result for (say) 3 o'clock tomorrow has to be taken with a pinch of salt, and this is where the human weather forecaster's experience and intuition comes in. She or he takes the computer's 'numerical product' as *guidance* – as a sort of framework. In the light of experience, he or she may decide to change, for example, how far a large area of rain is expected to move during the forecast period.

   I've said that there are 62 496 forecasting points at each level in the Met. Office's global computer model. But, irrespective of their precise number, *how are the points arranged*? In the Met. Office's global model they are placed at the intersections of circles of latitude and longitude. As you can see from Fig. 2, this leads to an uneven coverage of the globe: *there are many more points in a given area near the North and South Poles than there are in the same area near the equator.*

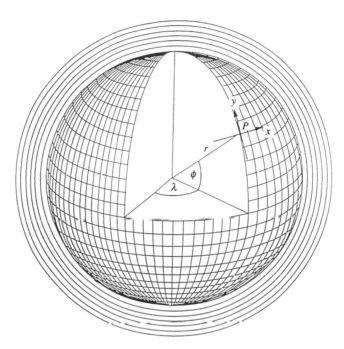

Fig. 2. Circles of latitude (ϕ) and longitude (λ) on a globe of radius *r*. In the UK Met. Office's global computer model, forecasting points are placed at the intersections of circles of latitude and longitude. (The concentric circles enclosing the globe are a qualitative indication of the many *layers* of points in the model.)

This uneven distribution leads to all sorts of difficulties, some of which you might not expect, and I'm not going to go into them. I will simply say that the Met. Office's global model takes various precautions which enable it to produce good forecasts in spite of its uneven distribution of forecasting points over the globe. But it is perfectly reasonable to wonder how things might be changed...

3. Designing a new computer model

Imagine that we are working on the design of a new computer model. We are unhappy about the uneven distribution of forecasting points in the present version, and we ask the key question:

How can we arrange our points evenly over the globe?

To collect our thoughts, let's imagine how we would answer the question if the Earth were flat (!) One way of achieving an even coverage of forecasting points

in the horizontal would then be to place them at the intersections of perpendicular lines, equally spaced (just like graph paper); see Fig. 3(a). This arrangement
is called a *square grid* of points. The 'grid unit' is a square. Each interior angle
of a square is 90°, of course.

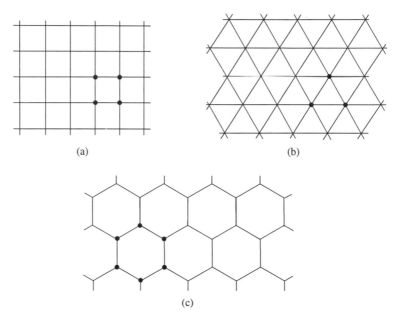

(a)   (b)

(c)

Fig. 3.   Regular polygonal grids on a plane: (a) a square grid, (b) an equilateral
triangular [ET] grid, and (c) a regular hexagonal grid. We place our forecasting points
where the lines cross ((a) and (b)) or meet ((c)); one 'grid unit' of points is indicated by
solid circles (•) in each case

Another solution is to use equilateral triangles as the grid unit; see Fig. 3(b).
*Equilateral triangles are going to figure so often in this article that I shall call them
ETs for short*. Each interior angle of an ET is 60°.

A further solution is a hexagonal pattern of points, as shown in Fig. 3(c). The
grid unit in this case is the regular hexagon – each interior angle of which
is 120°.

Let's limit attention to arranging our forecasting points on *regular polygonal
grids*. (This makes sure that the coverage of points has maximum symmetry.) We
have noticed three ways of doing this: with ETs, squares and (regular) hexagons.
Are there any other possibilities? To answer this question, let's get systematic.

4. Polygons: irregular and regular

Consider a polygon having N sides. Clearly it will also have N corners (or
vertices). The irregular polygon shown in Fig. 4 has 5 sides and 5 corners; it is
a pentagon.

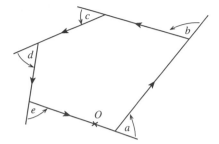

Fig. 4.   An irregular pentagon, showing its exterior angles *a, b, c, d, e*

The angles *a*, *b*, *c*, *d*, *e* are called the *exterior angles*. Suppose you start at point *O* and walk anticlockwise round the polygon (as indicated by the arrows with solid heads). When you get to a corner, you turn left by an angle equal to the appropriate exterior angle (as indicated by the other arrows). When you have returned to point *O* you have turned left by a total of 360 degrees. Hence

sum of exterior angles = *a* + *b* + *c* + *d* + *e* − 360 degrees

If the pentagon is *regular* – that is, if its sides are all of equal length – then the exterior angles *a*, *b*, *c*, *d*, *e* are each the same, and equal to 360/5 = 72°. The *interior angles* are each equal to 180 − 72 = 108°.

Clearly, a regular polygon having *N* sides (an '*N*-gon') has exterior angles of 360/*N* degrees. The interior angles, *I*, of the regular polygon are given by

$$I = 180 - (360/N) = 180\left(1 - \frac{2}{N}\right) \text{ degrees.}$$

(Putting *N* = 3, 4, 5, 6 in this equation gives *I* = 60°, 90°, 108°, 120°. These numbers agree with various results given earlier, so all is well.)

In Fig. 5 are shown regular *N*-gons for *N* = 3, 4, 5, 6, 7, 8, 9, 10 and 12, together with their names and their interior angles, *I*, obtained from our equation. We shall need most of this information later.

From now on, let's assume that the unit of angle is the degree. This will save a lot of tedious repetition of word or symbol. Mathematicians often use radians rather than degrees to measure angles, but degrees are often more convenient when regular figures are being examined. Notice how accommodating the number 360 is for us: it is divisible (without remainder) by 3, 4, 5, 6, 8, 9, 10 and 12 (as well as 2), so the only regular *N*-gons up to and including *N* = 12 which do not have whole-number interior angles are the regular heptagon (*N* = 7) and the regular hendecagon (*N* = 11).

Armed with these ideas, it's fairly easy to see that our square, ET and hexagon grids are the only ways of covering, or *tessellating*, the plane with regular polygons. At every 'corner' (where we are going to locate a grid-point) at least three polygons must meet. So the *maximum* interior angle is 120 – the hexagon value. Regular heptagons, and higher regular polygons, cannot completely cover the plane.

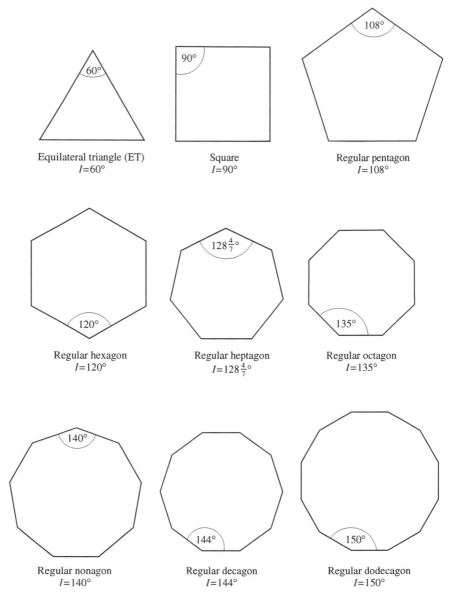

Fig. 5.   Some regular plane polygons: shapes, names and interior angles (I) in degrees

What about regular pentagons? As we have seen, they have interior angle $I = 180(1 - \frac{2}{5}) = 108$. Three of them can meet at a point, but there is an angle $360 - 324 = 36$ left over. Clearly, regular pentagons cannot tessellate the plane either.

So pentagons aren't relevant to the problem of covering the plane evenly. But that wasn't our original problem — we were really interested in gridding the sphere. We haven't seen the last of pentagons – not by a long chalk!

(By side-stepping the problem in hand – looking at regular distributions of points over a plane instead of over a sphere – we have gained some useful experience, and also come across something which will be very useful in our main problem. This is a common situation in mathematics and physics.)

5. The Platonic solids

Let's *cut out* the 36° angle left between our three pentagons, then *fold* along the lines joining the pentagons. Then, if we join the cut edges, we have made a three-dimensional structure which has a point, or *vertex*, where three pentagons meet; see Fig. 6.

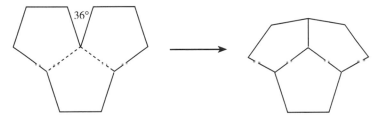

Fig. 6. Formation of a polyhedral angle from three regular pentagons

This structure is an example of a *polyhedral angle*. It obviously isn't any good for covering a flat plane, but it provides a way of covering a sphere evenly with points. The polyhedron known as the regular pentagonal dodecahedron (see Fig. 7) has 12 ('dodeca') faces, each of which is a regular pentagon. Each vertex, and each polyhedral angle – there are 20 of each – is the same in that each has the same symmetrical environment of neighbouring faces and vertices. To arrange 20 points regularly over a sphere we construct a regular pentagonal dodecahedron inside it, and put our points where the vertices touch the sphere (as indicated in Fig. 7). A polyhedron whose faces are all identical regular polygons is called a *regular polyhedron*, or a *regular solid*, or a *Platonic solid*.

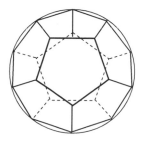

Fig. 7. A regular pentagonal dodecahedron – the 'dodo' – seen from above the midpoint of one of its faces. The circumscribing sphere, which touches each of the 20 vertices, is shown in outline

Classical Greek mathematicians knew of the existence of regular polyhedra and Plato was the first to describe them all. Pythagoras (over 100 years before Plato) seems to have missed one – the one you will be familiar with if you watch *Neighbours*.

Before asking you to try to find all of the regular polyhedra, it will be helpful if I introduce some abbreviated terminology and some notation.

Instead of referring to the 'regular pentagonal dodecahedron', let's simply call it the *dodo*. And let's call the polyhedral angle formed at a vertex where three regular pentagons meet, a $(5,5,5)$ polyhedral angle. Then, for example, a $(3,3,3,3)$ polyhedral angle is formed by the meeting of four equilateral triangles (ETs).

It will be helpful also to have a name for the amount, in degrees, by which the sum of the interior polygon angles, or *face angles*, at a vertex falls short of 360. We've already seen that this quantity is $360 - (3 \times 108) = 36$ for a dodo vertex. Let's call it the *curvature angle*, and denote it by the symbol C:

$$C = 360 - \text{sum of interior polygon angles (face angles) at vertex}$$

Suppose we try to make a polyhedral angle from four squares. We should already know that this won't work, because squares can cover (tessellate) the plane. Also, $360 - (4 \times 90) = 0$; i.e. the curvature angle, C, is 0. Clearly, the curvature angle of the constituent polygons in a polyhedral angle must be greater than zero.

Worksheet 1

1. Making a polyhedral angle with four squares won't work because they just cover the plane. Examine other ways of making polyhedral angles with squares. How many possibilities are there?
2. The dodo has three regular pentagons meeting at each vertex. Can you make any other polyhedral angles using only regular pentagons?
3. How many different polyhedral angles can you make using only equilateral triangles (ETs)?
4. Can you make a polyhedral angle out of regular hexagons?
5. Calculate the curvature angle, C, for each of the polyhedral angles that you have found in questions 1–4.
6. Now work out the quantity $V = (720/C)$ for each case (including the dodo). As we shall see later, V is the *number of vertices* in a regular polyhedron. What do the regular polyhedra which correspond to the polyhedral angles you have discovered in questions 1–4 look like?

You have, I hope, discovered five regular polyhedra as a result of doing Worksheet 1 (or as a result of looking at the Solutions). It is easy to see that there are no more than five regular polyhedra: we have shown that hexagons won't do – they just cover (tessellate) the plane – and all higher (regular) polygons have interior angles of more than 120°. It was Euclid who first showed that there are only five regular convex polyhedra. Diagrams of the cube and dodo have already been given (Figs 1(a) and 7); Fig. 8 shows the other three regular polyhedra.

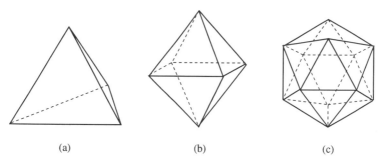

(a)   (b)   (c)

Fig. 8.   (a) The regular tetrahedron, (b) octahedron, and (c) icosahedron

A lot of things could be said about the Platonic solids, but I must limit myself to just three comments.

1. The five regular polyhedra represent extremes of symmetry, and many people find them interesting to contemplate. Over the ages, an almost mystic quality has been ascribed to them. In the late 16th century, Kepler proposed that the orbits of the planets about the Sun could be modelled by nested spheres with radii determined by the regular polyhedra. The conjecture worked remarkably well, in a quantitative sense, and Kepler felt he had discovered some underlying order which governed the planets. The whole notion was put into a harsher perspective by the discovery that the orbits of the planets about the Sun were generally not circular, but elliptical, and – more severely – by the existence of more than five planets.

2. It is quite easy to be more precise about the symmetry of the regular polyhedra. Consider the dodo. If you rotate it about an axis joining the midpoints of opposite faces, it will look the same five times in a complete revolution; it has five-fold symmetry about each such axis. It also has three-fold symmetry about each axis joining opposite vertices. Finally, it has two-fold symmetry about each axis joining the midpoints of opposite edges.

3. Suppose you are given one of those problems where you are asked to write down the next number in a series. For example:
$$0, 3, 8, 15, 24, 35, 48, 63, ?$$
This is easy enough – let's agree on 80. But what about this one:
$$4, 6, 8, 12, 20, ?$$
Don't think too hard. There is *no* next number in the series; these are the numbers of vertices (or, indeed, faces) of the regular polyhedra, and there are only five of them! If you find this rather annoying, I don't blame you, but before I promise not to offer any more red herrings, I'll tell you that you will soon know the next number in this series:
$$12, 24, 30, 48, 60, ?$$

But there is only one more number in the series; these are the numbers of vertices of the Archimedean solids (to be defined in Section 7). There are 13 such solids in total; mystics, please note!

6. Almost regular arrangements of points on a sphere

The (regular) tetrahedron, octahedron, icosahedron and dodecahedron have respectively 4, 6, 12 and 20 vertices; the cube has 8. An important result now follows.

It is impossible to arrange more than 20 *points regularly over a sphere.*

Constructing a regular icosahedron (instead of a dodecahedron) inside a sphere would give 12 points with identical symmetric environments on the sphere. Using a cube, a regular octahedron and a regular tetrahedron would give, respectively, 8, 6 and 4 points with identical environments on the sphere. But since there is no regular polyhedron with more than 20 vertices, we cannot arrange more than 20 points regularly over the sphere.

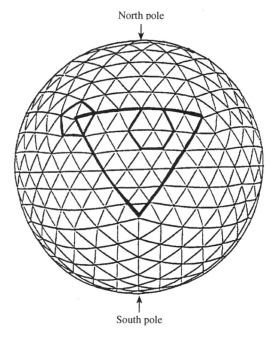

North pole

South pole

Fig. 9. A triangulated icosahedral grid on the sphere. Bold lines emphasize: one large triangular region (consisting of 36 smaller triangles), a typical environment of six neighbouring points (the hexagon), and a special environment of five neighbouring points (the pentagon). A total of 12 points – corresponding to the vertices of the inscribed icosahedron – have a special environment of five neighbours. (After Sadourny *et al.* (1968))

Our computer model design project has hit an obstacle! It is an insurmountable obstacle. We can get around it in several ways, but only at the expense of sacrificing our ideal of perfectly regular arrangement of points. One way is to triangulate the regular icosahedron, and to project the new points onto the circumscribing sphere. This achieves a much more even distribution of points than the latitude-longitude grid, but it is not regular: 12 points (at the vertices of the icosahedron) have five near neighbours, but all other points have six near neighbours. Fig. 9 tries to make this clear.

Another way of getting around the problem is to use polyhedra which have slightly lower symmetry than the Platonic solids: the *Archimedean* solids.

7. Archimedean solids: the principle

Every face of a Platonic solid is a regular polygon of one type: ETs for the (regular) tetrahedron, octahedron and icosahedron; squares for the cube; (regular) pentagons for the dodo. Every edge of a Platonic solid is the same length, and every polyhedral angle is the same.

As we have noted, the Platonic solids are also called the regular solids. Every edge of a *semi-regular* solid is the same length, every polyhedral angle is the same, and every face is a regular polygon, but different types of polygon are allowed: we can mix ETs, squares and other regular polygons so long as we do not violate the other requirements.

It turns out that there are three classes of semi-regular solids. Members of two of these classes – the regular prisms and regular antiprisms – are infinite in number; a brief description is given in the Appendix. The third class contains just 13 members: the *Archimedean solids* (which are much more interesting than the prisms and antiprisms).

Figure 10 shows my favourite Archimedian solid: the *snub cube*. (Your favourite Archimedean will probably be a different one, particularly if you are interested in football.)

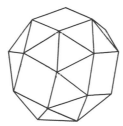

Fig. 10.   A snub cube (after Cundy and Rollett (1961))

The snub cube, which comes in left-hand and right-hand forms which are mirror images of one another, has six square faces and 18 ET faces. At each vertex, four ETs and one square meet. You can see that this is a viable polyhedral angle by adding up the constituent polygonal angles (just as we did earlier for the Platonic solids):

$$90 + 60 + 60 + 60 + 60 = 330 < 360$$

The curvature angle, C, is thus 30. If we apply our '720 rule' $[VC = 720]$ we find $V = 24$, which can easily be verified by counting the vertices of this figure.

Figure 11 depicts the formation of the snub cube's $(3,3,3,3,4)$ polyhedral angle by cutting out the 30° sector between four ETs and a square on a plane, and then folding.

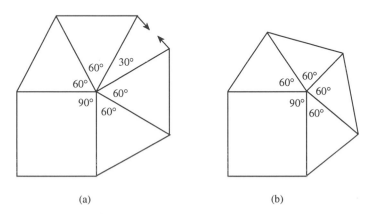

(a)   (b)

Fig. 11.   Formation of the polyhedral angle of the snub cube. (a) A square, four ETs and a 30° sector. (b) A 2-D diagram of the polyhedral angle formed by cutting out the 30° sector in (a), folding, and joining along the edges of the adjacent ETs (as indicated by the arrows in (a)). In our notation, the polyhedral angle is $(3,3,3,3,4)$

I think you will have guessed by now what our next exercise is going to be!

Worksheet 2

The summary information about regular plane polygons given in Fig. 5 may come in useful for this exercise.

1. What polyhedral angles can you make with square(s), regular hexagon(s) and regular octagon(s)? (Use at least one of each polygon.)
2. Repeat question 1 with square(s), regular hexagon(s) and regular decagon(s).
3. Repeat question 1 with ET(s) and regular octagon(s).
4. Repeat question 1 with ET(s) and regular decagon(s).
5. Try square(s) and regular hexagon(s).
6. Try ET(s) and regular hexagon(s).
7. Calculate the curvature angle, C, and $V = 720/C$, for each possible polyhedral angle that you have discovered in questions 1–6. Use *rough* sketches to investigate whether each polyhedral angle in question 1–6 can be combined with identical polyhedral angles to build up a closed figure.

8. As we have seen, the polyhedral angle $(3,3,3,3,4)$ forms the vertex of an Archimedean solid. What other polyhedral angles can you make with square(s) and ET(s)?

9. Repeat question 8 with regular pentagon(s) and regular hexagon(s).

10. Repeat question 8 with ET(s) and regular pentagon(s).

11. Try ET(s), square(s) and regular pentagon(s).

12. Calculate the curvature angle, C, and $V = 720/C$, for each possible polyhedral angle that you have discovered in questions 8–11, and examine the combination properties of each polyhedral angle (using rough sketches).

8. Archimedean solids: the practice

Here is a list of the polyhedral angles of the 13 Archimedean solids (together with brief comments for subsequent discussion):

$(3,6,6)$	Truncation of the vertices of a tetrahedron
$(4,6,6)$	Truncation of the vertices of an octahedron
$(3,8,8)$ $(3,4,3,4)$	Truncations of the vertices of a cube
$(4,6,8)$ $(3,4,4,4)$ $(3,3,3,3,4)$	Truncations of the vertices and edges of a cube
$(3,10,10)$ $(3,5,3,5)$	Truncations of the vertices of a dodecahedron
$(4,6,10)$ $(3,4,5,4)$ $(3,3,3,3,5)$	Truncations of the vertices and edges of a dodecahedron
$(5,6,6)$	Truncation of the vertices of an icosahedron

As the comments indicate, each of the corresponding Archimedean solids can be visualized as the result of applying careful surgery to a Platonic solid: judiciously cutting off ('truncating') the vertices, or vertices and edges. (In some cases, more than one visualization is possible, but I have not entered alternatives in the comments. Note that the qualification 'regular' has been omitted from the names of the Platonic solids.)

The visualizations are easily conveyed for the vertex truncations: $(3,6,6)$, $(4,6,6)$, $(3,8,8)$, $(3,4,3,4)$, $(3,10,10)$, $(3,5,3,5)$, $(5,6,6)$. See Fig. 12 for $(3,6,6)$ visualized as a truncation of a regular tetrahedron. The remaining cases, which involve truncations of edges as well as vertices, are more difficult to visualize, and it is really necessary to examine either the three-dimensional figures themselves or detailed diagrams of them.

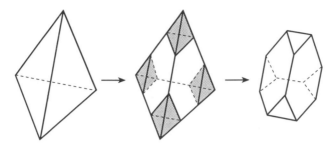

Fig. 12. Formation of the Archimedean solid having polyhedral angles (3,6,6). The vertices of the regular tetrahedron have to be cut off in just the right way to make the remaining faces *regular* hexagons. This formation may be more clearly demonstrated by truncating a paper or cardboard tetrahedron with scissors (the cuts being guided by straight lines drawn between neighbouring points of trisection of the sides of the original tetrahedron)

A different mode of visualization, which is not always applicable but is often easier than truncation, is *separation* of the faces of a Platonic solid to accommodate interposed squares or ETs. Figure 13 visualizes the Archimedean solid having polyhedral angles (3,4,4,4) as a separation (by 12 squares) of the six faces of a cube.

Because each polyhedral angle of an Archimedean solid is identical, each point on the sphere circumscribing and touching the solid has the same environment of points. But the arrangement of the points lacks the maximal symmetry of the arrangement obtained by using an inscribed Platonic solid instead. Also, we cannot achieve identical environments for more than 120 points (120 being the number of vertices in the Archimedean solid having (4,6,10) polyhedral angle). For these reasons, the Archimedean solids do not offer a serious solution to our computer model design problem; triangulation of the faces of the icosahedron, as described in Section 6, is the best near-solution.

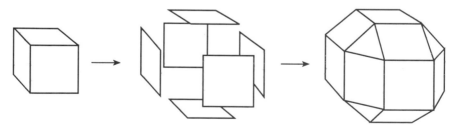

Fig. 13. Formation of the Archimedean solid having polyhedral angles (3,4,4,4). In the first step, the cube is *exploded*; the explosion is the very special one which leaves the six faces of the original cube just far enough apart to allow (second step) 12 identical squares to fit into the 12 gaps left by the parting of the original edges. There are eight triangular 'holes' opened up by this process (corresponding to the eight corners of the original cube), but these triangles are equilateral and have the same side as the squares, so the resulting 'separated' figure is Archimedean

9. Logic and visualization

Our treatment of the Archimedean solids is a good example of a *double approach* to problems posed by the physical world. You can approach a problem systematically (as we did in Worksheet 2), or you can use your eyes and imagination (as we have just done). In almost all problems in mathematical physics, either approach is possible (although in many cases the visualization will not be a direct geometric view of the problem). Using both approaches is a very powerful strategy.

It is remarkable how many mathematicians and physicists – even professional mathematicians and physicists – tend to use either one approach or the other, but not both. Mathematicians tend to adopt the systematic, step-by-step, procedural approach. Physicists tend to adopt the overview, gestalt, visual approach.

Felix Klein – a German mathematician who wrote a number of very readable books on mathematics as well as contributing greatly to its development in the late 19th century and early 20th century – observed that a similar division of approach occurred even amongst the geometers of his time (who were presumably all mathematicians). Much more recently, Malcolm Longair has commented on the same phenomenon amongst his fellow physicists.

If you take away only one lesson from this article, I would like it to be that the step-by-step and overview approaches to problems posed by the world around us are both equally valid, and *together* they offer a golden road to progress.

Another good reason for developing a visual approach to problems in mathematics and physics is that it helps us to communicate rationale and results to non-mathematicians and non-physicists. This is especially true of artists, architects, designers and engineers, who usually enjoy extremely good space perception.

10. Euler's formula

So far, we have concentrated on (convex) polyhedra which have a high degree of regularity and symmetry. Polyhedra which are much less regular might be expected to obey no particular rules, but this is not the case! If, for *any* convex polyhedron, you count:

the number of *faces*, F;

the number of *edges*, E;

and the number of *vertices*, V,

and you count correctly, the numbers will obey the following relation:

Theorem 1
$$F + V = E + 2$$
This result was discovered by Leonhart Euler, the great Swiss mathematician, in the 18th century. A proof (due to Cauchy) is not difficult; let's have a look at it.

Proof

Imagine the convex polyhedron to be built up face by face. This involves numbering the F faces $1, 2, 3, \ldots, F$ so that successive faces meet either in an edge or in a vertex. The early stages of an example are shown in Fig. 14.

Let V_i and E_i be the number of vertices and the number of edges accumulated when i faces have been assembled. The first face, since it is a polygon, has the same number of edges and vertices, so

$$E_1 - V_1 = 0.$$

Now add the second face. Whether this second face meets the first face in an edge or in a vertex, the net effect is to add to the totals one more edge than vertex. Thus

$$E_2 - V_2 = 1.$$

Now add the third face. Again, so long as it is connected – either via a vertex or a shared edge – to the growing polyhedron, the net effect is to add to the totals one more edge than vertex, so

$$E_3 - V_3 = 2.$$

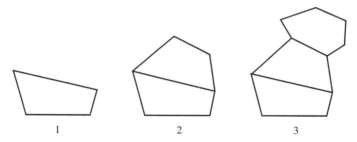

1   2   3

Fig. 14. The first three stages in assembling a polyhedron (as envisaged in Cauchy's proof of Euler's formula $F + V = E + 2$). In this example, two edges are shared between the first three faces

Every time we add a face – except the last face – we add one more edge than vertex:

$$E_i - V_i = i - 1 \qquad (0 < i < F).$$

When $i = F - 1$ (all faces added except the last) the running total of edges and the running total of vertices have both reached their final values E and V, since adding the last face will (obviously!) add one face, but *not* add a new edge or a new vertex. Finally, therefore:

$$E - V = F - 2$$

or

$$F + V = E + 2$$

which is Euler's Formula.   □

It is fun to verify Euler's Formula by counting the number of edges, faces and vertices of some specific polyhedra. If you make some fairly complicated polyhedra you may find, however, that Euler's Formula is acting as a check on your counting.

We can use Euler's Formula to prove the '720 rule' that we applied to polyhedral angles earlier. Here is the Theorem (which was discovered by Descartes) and a proof for the Platonic solids; you might like to investigate the case of the Archimedean solids yourself (and, indeed, the case of general convex polyhedra – for which some modification of the rule will be necessary).

Theorem 2

If V is the number of vertices of a Platonic or Archimedean solid, and C is the curvature angle of each polyhedral angle of the solid, then $VC = 720$.

Proof (for the Platonic solids)
Let N = number of sides of each polygonal face (i.e. the faces are N-gons)
and M = number of polygons that meet at each vertex.

If the Platonic solid has F faces, and E edges, then

$$NF = 2E \qquad (1)$$

since two *polygon* edges coincide in each edge of the solid.

Also $$NF = MV \qquad (2)$$

since: (i) M polygons meet at each vertex; (ii) there are V vertices; and (iii) the total number of polygonal angles (face angles) is NF.

So, from Euler's Formula in the form

$$2F + 2V - 2E = 4,$$

use of equation (1) shows that

$$2F + 2V - NF = 4;$$
$$\text{i.e.} \qquad 2V + F(2 - N) = 4. \qquad (3)$$

Use of equation (2) to substitute for F in equation (3) ($F = MV/N$) gives:

$$2V + \frac{M}{N}(2 - N)V = 4.$$

Re-arrange and multiply by 180:

$$V\left[360 - 180\left(1 - \frac{2}{N}\right)M\right] = 720.$$

The quantity in the square brackets is the *curvature angle*, C, of each vertex:

$$C = 360 - 180\left(1 - \frac{2}{N}\right)M.$$

So $$VC = 720 \qquad \text{which is the '720 rule'.} \qquad \square$$

I wanted to discuss Euler's Formula, if only briefly, for several reasons.

1. It has allowed us to prove the '720 rule' for the Platonic solids.
2. I didn't want us to discuss only highly regular and symmetrical polyhedra.

3. It is salutary to learn that even apparently irregular or non-symmetric systems obey rules. This often happens in the natural world. An extreme example from classical physics is the behaviour of gases: a typical sample of a gas consists of billions of molecules in random motion, but the whole sample will quite closely obey simple macroscale laws such as Boyle's Law (relating volume and pressure at constant temperature). You just have to look for rules in the right places, and learn to ask the right questions!

Solutions to Worksheet 1

1. Only three squares will work. Two squares would simply be back-to-back, and so would enclose no volume. A polyhedral angle must have at least three faces! The solid which results is the cube. In our notation the polyhedral angle formed by three squares is $(4, 4, 4)$. Note that $C = 360 - (3 \times 90) = 90$; thus $(720/C) = 8$, which is indeed the number of vertices of a cube.
2. No. $4 \times 108 = 432 > 360$.
3. Three: $(3, 3, 3)$; $(3, 3, 3, 3)$; $(3, 3, 3, 3, 3)$. Respective values of C and $720/C$ are: 180 and 4; 120 and 6; 60 and 12; regular tetrahedron, octahedron and icosahedron (as seen at the end of editions of *Neighbours*). $((3, 3, 3, 3, 3, 3)$ is one of the tessellations of the plane noted in Sections 3 and 4.)
4. No. $3 \times 120 = 360$. This is also one of the tessellations of the plane noted in Sections 3 and 4.
5. Covered in solutions to questions 1 and 3.
6. For the dodo, $C = 36$, so $V = 20$; the dodo has 20 vertices.

Solutions to Worksheet 2

You have probably discovered the polyhedral angles of a number of regular polygonal prisms and antiprisms as well as those of some Archimedean solids.

Regular polygonal prisms have polyhedral angle $(N, 4, 4)$, where N is any whole number greater than 2; regular polygonal antiprisms have polyhedral angle $(N, 3, 3, 3)$. These prisms form infinite classes of solids having identical polyhedral angles and regular polygonal faces; they are semi-regular, but not Archimedean (see Section 7 and the Appendix).

The Archimedean polyhedral angles, the curvature angles (C) and the numbers of vertices (V) corresponding to questions 1–6 and 8–11 are:

1. $(4, 6, 8)$; 15; 48

2. $(4, 6, 10)$; 6; 120

3. $(3, 8, 8)$; 30; 24

4. $(3, 10, 10)$; 12; 60

5. $(4, 6, 6)$; 30; 24

6. $(3, 6, 6)$; 60; 12

8. $(3, 4, 3, 4)$; 60; 12
   $(3, 4, 4, 4)$; 30; 24

9. $(5, 6, 6)$; 12; 60

10. $(3, 5, 3, 5)$; 24; 30
    $(3, 3, 3, 3, 5)$; 12; 60

11. $(3, 4, 5, 4)$; 12; 60

The 13th Archimedean polyhedral angle, $(3, 3, 3, 3, 4)$, was discussed in Section 7.

If you have discovered polyhedral angles which are not amongst these 13, and not prismatic or antiprismatic, they will represent structures which cannot be combined to form a closed solid. An example is $(5, 5, 6)$; this gives $C = 30$, but if you try to assemble five such angles at the vertices of a regular pentagon you end up with two hexagons wanting to share an edge (and hence a vertex)! You only have to draw a rough sketch to discover this.

The Archimedean solid having polyhedral angle $(5, 6, 6)$ projects onto its circumscribing sphere to give the pattern of some footballs (usually black pentagons and white hexagons).

Appendix: Prisms and antiprisms

Consider a polyhedral angle made up of the corners of two squares and any corner of any regular N-gon having the same side as each of the squares. The curvature angle of this polyhedral angle is:

$$C = 360 - (\quad 90 \quad + \quad 90 \quad + 180\left(1 - \frac{2}{N}\right)) = \frac{2}{N} > 0$$

$$square + square + polygon$$

This implies that any regular polygon will make a possible polyhedral angle with two squares. Indeed it will, and a figure having equal polyhedral angles results from replicating the angle around the polygon (by adding $N - 2$ squares) and then placing a second N-gon opposite the first. The figure is called a regular polygonal prism; it has polygon faces top and bottom (say) and squares around the side. An example is shown in Fig. 15(a).

A second infinite class of semi-regular solids has each polyhedral angle made up of three ET angles and any regular polygon angle:

$$C = 360 - (\quad 60 + 60 + 60 + 180\left(1 - \frac{2}{N}\right)) = \frac{2}{N} > 0$$

$$ET + ET + ET + polygon$$

The figure in this case is called a regular polygonal antiprism. It has polygonal

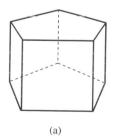

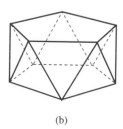

(a)   (b)

Fig. 15. (a) A regular pentagonal prism and (b) a regular pentagonal antiprism

faces 'top' and 'bottom', and $2N$ ET faces around the side; the upper polygon is rotated about a 'vertical' axis by $180/N$ degrees with respect to the lower one. Figure 15(b) shows an example.

Notice that a cube is a regular square prism and that a regular octahedron is a regular triangular antiprism.

Regular prisms and antiprisms may be considered as 'separations' (by squares and ETs respectively) of a regular N-gon seen as a plane figure having two faces. Separations of the Platonic solids were discussed in Section 8 in connection with the Archimedeans; it is interesting to note that a regular N-gon may be regarded as a Platonic solid in all but volume.

Acknowledgements

Figure 9 is adapted from the paper by R. Sadourny, A. Arakawa and Y. Mintz referenced below. It is reproduced by permission of the American Meteorological Society.

Figure 10, adapted from H. M. Cundy and A. P. Rollett's book *Mathematical Models* (also referenced below), is reproduced by permission of Oxford University Press.

Figure 2 is adapted from Met. Office College notes, written by Mr Eddy Carroll, on 'Numerical Weather Prediction' and is reproduced by permission of the Principal.

References

H. M. Cundy and A. P. Rollett, *Mathematical Models* (OUP, 1961)

This is a classic text on polyhedra (amongst other subjects). It presents a wealth of information about the Archimedean and Platonic solids and related figures. It contains many diagrams, including nets for the construction of polyhedra. These nets are for copying, not for cutting out of the book! If you want ready-to-cut nets for polyhedra then several books by Gerald Jenkins and Anne Wild will interest you. Examples are the three slim volumes of *Make Shapes – Mathematical Models* (ISBN 0-906212-014; Tarquin Publications, Diss, Norfolk.)

R. Sadourny, A. Arakawa and Y. Mintz, 'Integration of the non-divergent barotropic vorticity equation with an icosahedral-hexagonal grid for the sphere', *Monthly Weather Review*, **96** (1968) 351–356

If the best solution to our computer model design problem (see main text) was proposed in 1968, why is it not being used for global weather forecasting now? In my view, it deserves more attention than it has received; but the answer is that geometric regularity is not the only consideration. Mathematical and computational convenience are at least as important in practice, and they have compelled other choices. However, this situation may be changing.

Further Reading

J. Daintith and R. D. Nelson, *The Penguin Dictionary of Mathematics* (Penguin Books, London, 1989)

Dictionaries are less likely than specialized texts to overwhelm you with information, but they must be comprehensible and correct (by and large). Of the three mathematical dictionaries of which I am aware, I think this one strikes the best general balance and it is enthusiastically good on polyhedra.

F. Klein, *Geometry* Translated by E. R. Hedrick and C. A. Noble (Dover, New York, 1939)

Klein's observation about geometers (cited here in Section 9) comes from this book, which is fairly advanced (more so than the uniform *Arithmetic, Algebra, Analysis* – an equally laudable text). The author's perspective and wit make for compelling reading, and the mathematical insights are deep.

M. S. Longair, *Theoretical Concepts in Physics* (CUP, 1984)

Longair's book – also cited here in Section 9 gives a number of striking historical surveys (including an account of Kepler's work). It shows very clearly how mathematicians and physicists, with their typically different approaches to scientific problems, have together contributed to modern understanding of the world around us.

E. N. Lorenz, *The Essence of Chaos* (UCL Press, London, 1993)

I recommend Chapter 3 as an introduction to weather forecasting: it is a concise and lucid description in a fertile context. The book benefits from the author's dry humour as well as from his authority and clarity.

D. Wells, *The Penguin Dictionary of Curious and Interesting Numbers* (Penguin Books, London, 1983)

This is a brilliant book which discusses the Platonic and Archimedean solids (under 5 and 13, of course!) – and much, much else.

12 Water Waves

by WINIFRED WOOD
Department of Mathematics, University of Reading

1. Introduction

Imagine you are standing by a pond on a still day. There is not a breath of wind and the water surface is absolutely flat. You pick up a large stone and heave it into the water. You know what will happen – the stone makes a splash at the place where it enters the water and then small waves travel out from that place in concentric circles as in Fig. 1.

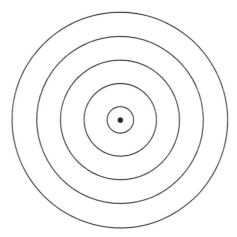

Fig. 1. Concentric circles

The waves continue to travel out for a while and then they die away. The pond looks as it did before the stone was thrown in. What do we mean when we say the waves travel out from where the stone went in? What travels? The pond looks exactly as it did before. The water has not gone anywhere. The wave effect is a disturbance in which the particles of water move only locally but the disturbance at one position causes the disturbance at the next position and thus the wave travels.

The effect is rather like shunting trucks. The end truck is pushed, it knocks against the next one, that causes the next in line to move and so on. Each truck moves only a short distance but the movement is transmitted along the line.

You can see something similar when you are in a car stuck in a long traffic jam on a crowded motorway, for example, reduced to a single lane by road works. You see a car in the distance move a little way and you know that the effect will be transmitted along the line until it reaches your car, your car moves, then the one behind and so on. The effect is like a wave travelling along the line of cars – a small local disturbance continues to be transmitted until it reaches the end of the line.

Out in the ocean the disturbance is usually caused by the strong winds in violent storms. The strong winds are produced by the weather systems and

Fig. 2. Storm waves developing

Fig. 3a. Long-crested waves on the Pacific Ocean

Fig. 3b.   Long-crested waves approaching a straight beach

because the water is so mobile a lot of energy passes from the wind to the waves. The longer the wind blows the higher the waves. In the middle of the storm there is a confusion of waves as in Fig. 2. We shall find that as the waves travel out from the storm they sort themselves out into a more regular pattern. At some distance from the storm they can form a steady train of long-crested waves as in Fig. 3a and the background of Fig. 3b. The crest is the highest point of the wave and 'long-crested' implies they are moving at right angles to their crests so that these remain straight and parallel.

If you look at a floating seagull in deep water as a steady train of these long-crested waves goes by you will see that it moves round in a vertical circle. The local movement of a particle of water on the surface of the sea here is a circle – call the radius of the circle a. As the wave goes by from right to left (Fig. 4) the particle moves clockwise in this circle, radius a, with its centre O on the mean water level (MWL). This highest point of the wave is called the crest and

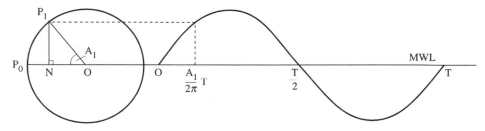

Fig. 4.   Circle and profile for deep water waves

the lowest the trough. We can trace the profile of the wave by tracing the change in height of the particle as it moves round the circle starting from the point P_O on the MWL. At some later time the particle is at P_1 where the angle $P_O\hat{O}P_1$ is A_1 (Fig. 4). It is convenient here to measure this angle in radians. Before introducing radians we look first at the measurement of angles in degrees. One whole rotation is 360°.

(Why the number 360? Why not 400? The ancient Babylonians thought there was something special about the number 60 and also about equilateral triangles – hence 60° for the measure of an angle in an equilateral triangle and 360° for a complete rotation. The number 360 is also an approximation for the number of days in a year.)

In radian measure one complete rotation is 2π radians

i.e. $360° = 2\pi$ radians.

The number π is so important it is on every pocket calculator to the number of digits that the calculator can carry, 3.14159.... . The circumference of the circle of radius u is $2\pi u$, which is the radius times one complete rotation measured in radians. Hence the length of the arc $P_O P_1$ in Fig. 4 is the radius a times A_1 (where A_1 is measured in radians).

Suppose the particle on the surface of the sea takes a time T to travel once all round the circle – this is the time for one complete wave to travel by and it is called the period of the wave.

The speed of the particle is

$$\frac{\text{distance}}{\text{time}} = \frac{2\pi a}{T}$$

and the time it takes to get from P_O to P_1 is $A_1 T/2\pi$. If P_1 is halfway between P_O and the highest point P_2

$$P_O\hat{O}P_1 = \frac{\pi}{4}$$

and the time the particle takes to travel from P_O to P_1 is $T/8$.

When the particle is at P_1 the height of the particle above MWL is $P_1 N$ (Fig. 4) where $P_1 N$ makes a right angle (90° = $\pi/2$ radians) with OP_O. Figure 5 shows

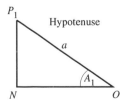

Fig. 5.   Right-angled triangle

the right-angled triangle P_1NO drawn separately. The longest side (the radius of the circle) is called the hypotenuse – this is opposite the right angle. The ratio

$$\frac{P_1N}{OP} = \frac{P_1N}{a}$$

is called the sine of the angle $P_0\hat{O}P_1 = A_1$. This is written as $\sin A_1$,

i.e. $$\sin A_1 = \frac{\text{length of side opposite the angle } A_1}{\text{hypotenuse}}.$$

Thus in general if we start measuring time from when the particle is at P_O on the MWL, the height of the particle above the MWL is

$$a \sin A$$

at the time $AT/2\pi$ and we can plot the profile of the wave by finding the heights for different values of A as the particle P moves around the circle. For example Fig. 6 shows an equilateral triangle with one angle bisected which

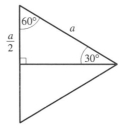

Fig. 6.   Equilateral triangle

should make it clear at what time the particle is at height $a/2$. Also $A = \pi/4$ at time $t = T/8$ and at that time the height of the particle above MWL is $a \sin(\pi/4)$. One of the examples on Worksheet 1 shows how to find out what this is.

You can see that the particle will again be at the same height at time $t = 3T/8$ and at the same distance below the MWL at $t = 5T/8$ and $t = 7T/8$. With just a few values for the sine of an angle you can trace the profile of the wave. This is called a sine wave because of its connection with the sine of the angle.

There are two ways of looking at a wave. You can either fix your eyes on a particular point in space and see a wave go by in a time T, i.e. the period T is the time observed for the passage from one crest to the next. Alternatively you can take a photograph of the sea at a particular point in time and see the length

Fig. 7.   Profile of wave with length marked

L from one crest to the next in space as shown in the profile in Fig. 7. The maximum displacement of the particle on the surface from MWL as the wave goes by, i.e. the radius of the circle which we have called a, is the amplitude of the wave. Note that for this sine wave theory to apply here we must have

$$a \ll L$$

which is shorthand for

a is small compared with L.

Approximately the theory applies when

$$a < L/14$$

i.e. a is less than $L/14$. (You can see that in Fig. 7 the horizontal and vertical scales must be different). As the amplitude approaches this value the wave becomes unstable, develops a crest and breaks.

Worksheet 1

1. 2π radians $= 360°$. What are $\dfrac{\pi}{6}, \dfrac{\pi}{4}, \dfrac{\pi}{3}, \dfrac{\pi}{2}, \dfrac{2\pi}{3}$ radians in degrees?
2. When the particle has travelled from P_0 to P_1 such that its height above MWL is $a/2$ the angle $P_0\hat{O}P_1 = \pi/6$. Show that $\sin(\pi/6) = 0.5$.

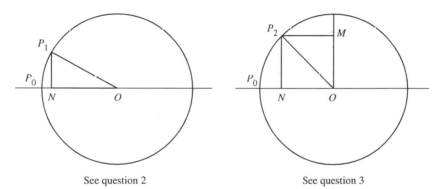

See question 2   See question 3

3. When the particle has travelled half way round the first quadrant of the circle to the point P_2, P_2MON is a square. What is the angle $P_0\hat{O}P_2$?

   Pythagoras' theorem says that with any right-angled triangle with sides of length a, b, c, with a opposite the right angle (the right angle $= 90°$),

   $$a^2 = b^2 + c^2$$

   Use this to find $\sin\left(\dfrac{\pi}{4}\right)$.
4. Now find $\sin\left(\dfrac{\pi}{3}\right)$.

   Check your results by using your calculator but remember to check whether it is using degrees or radians.

5. Use squared paper to plot the points on the sine wave when the particle has moved a distance $a(\pi/6)$, $a(\pi/4)$, $a(\pi/3)$, $a(\pi/2)$, round the circle and all the other points you can get with these values, marking in the corresponding times.

2. Waves in deep water

We are considering here waves in deep water which means that the depth is greater than the wave length L. In deep water the particles moves in circles which get smaller with the depth so that below a depth L the movement is

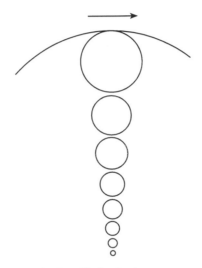

Fig. 8.   Circles in deep water

'negligible' (Fig. 8). Then the theory of waves in deep water is found to give the formulae

$$\text{wave speed, } c = \frac{L}{T} = \frac{g}{2\pi}T, \text{ i.e. } L = \frac{g}{2\pi}T^2 \qquad (1)$$

$$\text{which also gives } c^2 = \frac{g}{2\pi}L \qquad (2)$$

where g is the acceleration due to gravity. These sine waves are also called surface gravity waves. The latter 'c' here stands for celerity which also means speed; 's' is used for other things. Note that it is important to choose these symbols with care to avoid confusion.

You can see from the formulas above that a particular property of waves in deep water is that the longer the period T the faster the waves travel, and the longer the wave length L the faster the waves travel.

For example, if we use metres and seconds (which naturally fit in here because typically sea waves can have lengths of 1 m to 1000 m and periods 0.8 to 25 seconds) we can take

$$\frac{g}{2\pi} \approx (1.25)^2$$

('≈' means 'approximately' we do not have to be very accurate here).

Then

$$c = 1.25 \sqrt{L}, \qquad T = 0.8 \sqrt{L}, \qquad c = (1.25)^2 T.$$

(Check that these are consistent with the formulas above.)

For $L = 1$ m, $c = 1.25$ m per second, $T = 0.8$ second

$L = 100$ m, $c = 12.5$ m per second, $T = 8$ seconds

$L = 1000$ m, $c = 40$ m per second, $T = 25$ seconds

(but remember this 1000 m wave must be in an ocean with depth greater than 1 km for the deep water theory to apply).

3. Waves in shallow water

As the waves travel out from the storm area where they are generated, the waves with the longer periods and longer lengths travel faster. Hence long waves arriving on the coast from a large ocean can mean that a storm is approaching. In smaller enclosed seas the waves cannot travel far enough to sort themselves out and tend to arrive at the coast in a mixture of long and short waves. As the waves approach the coast the sea becomes shallower and they are no longer deep water waves. The particles now move in ellipses which get thinner near the bottom as shown in Fig. 9. The period T stays the same but when the depth of

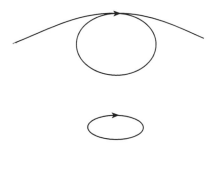

Fig. 9.   Ellipses in shallow water

Fig. 10.   Waves approaching the shore of Cornwall

Fig. 11.   Waves approaching the bay of Scarborough

the water h is small compared with the wave length L ($h \ll L$), the wave is called a long wave and the speed is found to be given by

$$c^2 = gh \tag{3}$$

$$\text{and} \quad L = cT = \sqrt{gh}\ T \tag{4}$$

i.e. the speed now simply depends on the depth. This means that waves of different periods travel with the same speed near the coast. For example, an 8-second wave which was 100 m long and travelling at $c = 12.5$ m per second in deep water becomes, at a depth of $h = 1$ m, a wave of 25 m travelling at 3.1 m per second. Thus the waves become shorter and travel more slowly as they approach the coast and this is why they align themselves with the coast as shown in Figs. 10 and 11. Figure 10 shows waves aligning themselves with part of the coast of Cornwall and Fig. 11 shows waves approaching and taking up the shape of the bay at Scarborough.

4. Groups of waves

In the first section we looked at a single sine wave. A violent storm at sea can be regarded as generating a confused collection of sine waves. It is only as they travel out from the storm area that they sort themselves out to some extent because the longer waves travel faster. The idea that the confused pattern of waves in the storm area can be seen as a sum of simple sine waves is very important and can be generalized to many other situations. (This called Fourier analysis).

As the waves travel out from the storm area they do not sort themselves out into neat packets of 20-second, 19-second, ... waves. They travel in groups of nearly the same period and wave length gradually spreading out. If you stand on the sea shore and look at the waves coming in you can see that they are not all the same height. There is a big wave and then four or five or perhaps seven smaller waves and then another big one. This is an effect produced by combining sine waves of nearly equal length so that they sometimes reinforce and sometimes cancel each other. Figures 12, 13 and 14 illustrate this with wave lengths as shown. In some parts of the world you will hear people say that it is every seventh wave that is highest, in others that is the third or the ninth. In Figs. 12, 13 and 14 there are the results of adding two sine waves which give every 6th, 11th and 7th waves largest.

Suppose there are two waves with wave lengths L_1 and L_2, with L_2 slightly bigger than L_1. At a particular moment in time you have a picture of the combined effect of these two waves (Fig. 12). The crests are together at the point P and again at Q where

$$nL_1 = (n - 1)L_2$$

where n is an integer. (In Fig. 14, $n = 6$)

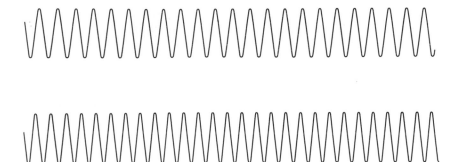

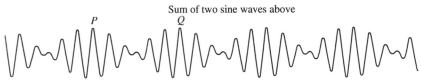

Sum of two sine waves above

Fig. 12.   Two sine waves combined, $L_1 = 10$, $L_2 = 12$ with P, Q marked

This means that

$$nL_1 = nL_2 - L_2, \quad \text{i.e.} \quad n(L_2 - L_1) = L_2$$

i.e.

$$n = \frac{L_2}{L_2 - L_1}.$$

Sum of two sine waves above

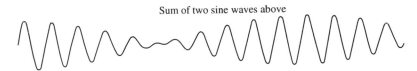

Fig. 13.   Two sine waves combined, $L_1 = 10$, $L_2 = 11$

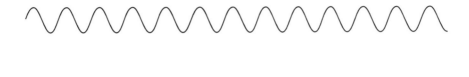

Sum of two sine waves above

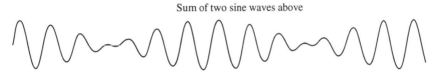

Fig. 14. Two sine waves combined, $L_1 = 6$, $L_2 = 7$

So these two sine waves combine to make a 'wave group' whose distance from crest to crest is

$$L = nL_1 = \frac{L_1 L_2}{L_2 - L_1}. \tag{5}$$

In shallow water the waves are all travelling at the same speed $c = \sqrt{gh}$ dependent on the depth h. So the group travels with this speed also and when you are standing at the edge of the sea you see the waves of the group travelling with the same speed.

In deep water the speeds of the two waves of different length are different because, from equations (1) and (2),

$$c = (1.25)^2 T = 1.25 \sqrt{L}$$

(using metres and seconds as before).

So now the waves in the group are travelling with slightly different speeds and we show in the next section that the group travels with speed approximately $c/2$, where c is the average speed in the group. This means that the individual waves are travelling through the group and if you try to follow a wave with your eyes you will find it disappears and another appears at the other end. Remember we began this chapter by talking about the effect of heaving a large stone into a pond. It is likely that the waves produced there will be short enough to be 'waves in deep water' and you will be able to see the longer waves on the outside of the disturbance and the group effect of waves appearing and disappearing.

5. Group speed

With a group of two waves the group wave length from equation (5) is

$$L = \frac{L_1 L_2}{L_2 - L_1}$$

If we put wave number k = the reciprocal of the wave length,

i.e.
$$k_1 = \frac{1}{L_1}, \quad k_2 = \frac{1}{L_2}$$

then the group wave length from equation (5), dividing top and bottom by $L_1 L_2$, is

$$L = \frac{1}{k_1 - k_2}. \tag{6}$$

The time between the occasions when the crests combine is

$$nT_1 = (n - 1)T_2$$

i.e.
$$n = \frac{T_2}{T_2 - T_1}$$

and the group period is

$$T = nT_1 = \frac{T_1 T_2}{T_2 - T_1}. \tag{7}$$

Put the frequency f = the reciprocal of the period,

i.e.
$$f_1 = \frac{1}{T_1}, \quad f_2 = \frac{1}{T_2},$$

then the group period

$$T = \frac{1}{f_1 - f_2}. \tag{8}$$

Hence from equations (6) and (8) the group speed

$$C = \frac{\text{group length}}{\text{group period}} = \frac{f_1 - f_2}{k_1 - k_2} = \frac{\text{difference in the frequencies}}{\text{difference in the wave numbers}}. \tag{9}$$

We want to show that when the two waves are of nearly the same period the group speed is approximately $c/2$ where c is the average speed of the two waves – this is done with the following algebra.

We first express the group speed using equations (5) and (7) as

$$C = \frac{L_1 L_2}{(L_2 - L_1)} \frac{(T_2 - T_1)}{T_1 T_2}.$$

Next substitute $$L_1 = \frac{g}{2\pi} T_1^2, \quad L_2 = \frac{g}{2\pi} T_2^2$$

from equation (1).

This gives $$C = \frac{g}{2\pi} \frac{T_1^2 T_2^2}{(T_2^2 - T_1^2)} \frac{(T_2 - T_1)}{T_1 T_2}.$$

Check that $(T_2 - T_1)(T_2 + T_1) = T_2^2 - T_1^2$ (this is called 'the difference of two squares') and cancel to give

$$C = \frac{g}{2\pi} \frac{T_1 T_2}{(T_2 + T_1)}. \tag{10}$$

We want to show that if $T_2 = T_1 + d$ where $d \ll T_1$ (i.e. d is small compared with T_1), then C is approximately half the average speed of the two waves. We do this by showing that the fractional difference between C and $(c_1 + c_2)/4$ is small. The fractional difference is

$$\frac{(c_1 + c_2)/4 - C}{(c_1 + c_2)/4} = 1 - \frac{4C}{c_1 + c_2} = 1 - \frac{4T_1 T_2}{(T_1 + T_2)^2},$$

substituting from equations (1) and (10).

Hence the fractional difference is

$$\frac{(T_1 + T_2)^2 - 4T_1 T_2}{(T_1 + T_2)^2}.$$

Check that $$(T_1 + T_2)^2 = T_1^2 + 2T_1 T_2 + T_2^2$$

and $$(T_2 - T_1)^2 = T_2^2 - 2T_1 T_2 + T_1^2.$$

Then the fractional difference is

$$\frac{(T_2 - T_1)^2}{(T_1 + T_2)^2} = \frac{d^2}{(2T_1 + d)^2}.$$

But since we have said that d/T_1 is small, $(d/T_1)^2$ is even smaller and we have shown that

$$C \approx \tfrac{1}{4}(c_1 + c_2) = \text{half the average speed of the two waves.}$$

For those familiar with differential calculus, the theory can be applied to waves whose lengths and periods are changing continuously and the group speed is then shown to be

$$\frac{df}{dk} = \tfrac{1}{2}c.$$

Note that some textbooks take the frequency and wave number as defined here, but others take

$$\text{frequency} = \frac{2\pi}{\text{period}}$$

and

$$\text{wave number} = \frac{2\pi}{\text{wave length}},$$

but you can see that since the group speed comes from a ratio of these the result is the same.

Worksheet 2

1. Use a pocket calculator to check the numbers used so far with $g = 9.81$ metres per second per second and π as given by the calculator. Notice that the numbers have been 'rounded off' to two figures as this is sufficiently accurate here.

2. In Fig. 12, $L_2 = 12$ and $L_1 = 10$ so $n = 6$. Check the values of n for Figs. 13 and 14.

   Taking the lengths of the waves in these figures to be given in metres, find their deep water speeds and the corresponding group speeds for these three cases.

3. In the Introduction we defined the sine of an angle in a right-angled triangle, i.e. an angle less than $\pi/2$ radians.

   Note that from the sine curve you can extend this idea to include the sine of an angle $P_0\hat{O}P$ where P is anywhere on the circle in Fig. 4. Draw a circle and mark in roughly the positions of P for this angle to be

   (i) $\dfrac{3\pi}{4}$ (ii) $\dfrac{2\pi}{3}$ (iii) π (iv) $\dfrac{7\pi}{6}$ (v) $\dfrac{3\pi}{2}$ (vi) 2π.

   Find the sines of these angles.

4. Use the exponential button on your calculator to work out some values of $(e^x - e^{-x})/(e^x + e^{-x})$, which is the definition of the hyperbolic tangent of x (written $\tanh x$).

   Plot a graph for $-10 < x < 10$.

6. Other sea waves

In the first two sections of this chapter we used sine waves to represent sea waves. We had one formula for the speed of waves in deep water and another

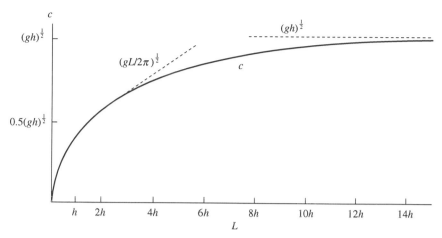

Fig. 15. The wave speed c for waves of length L in water of depth h. This shows the transition between the deep water value and the long wave value

formula for the speed in shallow water and you may wonder what happens in between. This is represented by the square root of a function called the 'hyperbolic tangent' (see Worksheet 2, question 4). Figure 15 shows how this curve joins the deep water wave theory with the shallow water wave theory. Mathematics is rather like a large jigsaw puzzle. You have a few pieces of the pattern here and a few pieces there and one day you realize how they join up and then you have made a great discovery.

Waves generated by the strong winds in storms at sea have periods of up to about 20 seconds corresponding to deep water waves of length about 624 metres travelling at a speed of about 31 metres per second. Much longer and faster waves can be generated by earthquakes which dislocate the earth's crust or cause large landslides at the bottom of the ocean. We expect them to be long waves, i.e. shallow water waves, because they are generated by a shock wave which travels through the whole depth of the ocean. They can have a period of around 16 minutes and travel at 150–200 metres per second. They are so long that, except perhaps in the deepest part of the ocean, they are shallow water waves. When these waves approach the coast the period stays the same but they slow down (speed = $\sqrt{gh}$) and shorten ($L = \sqrt{gh}\,T$) so that they become steeper. These earthquake waves have a small amplitude when they are travelling across the ocean but when they run up the coast they are still travelling quite fast and they become very high and can do a lot of damage. These earthquake waves are often called by the Japanese name tsunami.

The tides are also very long waves, so long that they are effectively shallow water waves. The tides are the combined effect of the gravitational pull of the Moon and the Sun on the seas around the Earth. The Moon goes round the Earth and the Earth and the Moon together go round the Sun. The predominant effect at the latitude of the UK is the tide from the gravitational pull of the

Moon with a period of 12.4 hours which is the familiar twice-a-day tide. The interaction with the influence of the Sun produces an effect like that shown in Figs. 12, 13 and 14. The two sometimes combine when the Sun, Moon and Earth are more or less in line, to give high tides (springs) and sometimes cancel out to give low tides (neaps). It is interesting to observe that in some estuaries the tide comes in (the flood) noticeably faster than it recedes (the ebb). This is because, as a shallow water wave, the speed is $\sqrt{gh}$ so the crest is moving faster than the trough.

The sine wave theory that we have been using is very useful but it only applies so long as the amplitude of the wave is small compared with its length. We can use these simple wave theories for many practical problems connected with the sea. There are more complicated theories for use for problems such as the wave forces on offshore platforms. Also there is still research to be done on the theory of what happens when waves become unstable and break.

Solutions to Worksheet 1

1. $\dfrac{\pi}{6} = 30°$, $\quad \dfrac{\pi}{4} = 45°$, $\quad \dfrac{\pi}{3} = 60°$, $\quad \dfrac{\pi}{2} = 90°$, $\quad \dfrac{2\pi}{3} = 120°$.

3. $P_O\hat{O}P_2 = \dfrac{\pi}{4} = 45°$, $\quad \sin(45°) = 0.707$ to 3 figures.

4. $\sin\left(\dfrac{\pi}{3}\right) = 0.866$ to 3 figures.

Solutions to Worksheet 2

2. In Fig. 13 $n = 11$. In Fig. 14 $n = 7$.

   The waves have deep water speeds $= 1.25\sqrt{L}$.

   In Fig. 12 speeds are 3.95, 4.33 m per second.

   The group speed is half the average $= \frac{1}{4}(3.95 + 4.33) = 2.07$ m per second.

   In Fig. 13 the speeds are 3.95, 4.14 m per second and the group speed is 2.02 m per second.

   In Fig. 16 the speeds are 3.06, 3.31 m per second and the group speed is 1.59 m per second.

3. (i) $\sin\left(\dfrac{3\pi}{4}\right) = \sin\left(\dfrac{\pi}{4}\right) = 0.707$, $\qquad$ (ii) $\sin\left(\dfrac{2\pi}{3}\right) = \sin\left(\dfrac{\pi}{3}\right) = 0.866$,

   (iii) $\sin(\pi) = 0$, $\qquad$ (iv) $\sin\left(\dfrac{7\pi}{6}\right) = -\sin\left(\dfrac{\pi}{6}\right) = -0.5$,

   (v) $\sin\left(\dfrac{3\pi}{2}\right) = -1$, $\qquad$ (vi) $\sin(2\pi) = 0$.

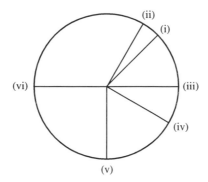

4.

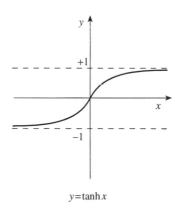

$y = \tanh x$

Acknowledgements

Copyright permission is gratefully acknowledged from the following sources.

Fig. 2   Institute of Oceanographic Sciences, Southampton, U.K.

Fig. 3b   Dr C. A. Fleming, Halcrow, Consulting Engineers.

Fig. 12   Cambridge University Collection of Air Photographs.

Fig. 13   B. Goss, 4 Langthorne Crescent, Grays, Essex.

Fig. 17   M. J. Lighthill, *Waves in Fluids* (Cambridge University Press, 1978).

Fig. 3 is a U.S. Army Air Corps official photograph (1933).